X-RAY DIFFRACTION TOPOGRAPHY

Other titles in the International Series in
THE SCIENCE OF THE SOLID STATE

Vol. 1: GREENWAY AND HARBEKE: Optical Properties and Band
Structures of Semiconductors

Vol. 2: RAY: II-VI Compounds

Vol. 3: NAG: Theory of Electrical Transport in Semiconductors

Vol. 4: JARZEBSKI: Oxide Semiconductors

Vol. 5: SHARMA AND PUROHIT: Semiconductor Heterojunctions

Vol. 6: PAMPLIN: Crystal Growth

Vol. 7: SHAY AND WERNICK: Ternary Chalcopyrite Semiconductors.
Growth, Electronic Properties and Applications

Vol. 8: BASSANI AND PASTORI PARRAVICINI: Electronic States
and Optical Transitions in Solids

Vol. 9: SUCHET: Electrical Conduction in Solid Materials

Forthcoming titles in the series

HOLT and HANEMAN: Defects and Surfaces in Semiconducting
Materials and Devices

ROY: Tunnelling and Negative Resistance Phenomena in p-n
Junctions

WILLIAMS: Luminescence and the Light Emitting Diode

X-RAY DIFFRACTION TOPOGRAPHY

by

B. K. TANNER, M.A., D. Phil.
Lecturer in Physics, Durham University

PERGAMON PRESS

OXFORD · NEW YORK · TORONTO · SYDNEY · PARIS · FRANKFURT

U. K.	Pergamon Press Ltd., Headington Hill Hall, Oxford OX3 0BW, England
U. S. A.	Pergamon Press Inc., Maxwell House, Fairview Park, Elmsford, New York 10523, U.S.A.
C A N A D A	Pergamon of Canada Ltd., P.O. Box 9600, Don Mills M3C 2T9, Ontario, Canada
A U S T R A L I A	Pergamon Press (Aust.) Pty. Ltd., 19a Boundary Street, Rushcutters Bay, N.S.W. 2011, Australia
F R A N C E	Pergamon Press SARL, 24 rue des Ecoles, 75240 Paris, Cedex 05, France
W E S T G E R M A N Y	Pergamon Press GmbH, 6242 Kronberg-Taunus, Pferdstrasse 1, Frankfurt-am-Main, West Germany

First edition 1976

Library of Congress Cataloging in Publication Data

Tanner, Brian Keith.
X-ray diffraction topography.

(I S science of the solid state; v. 10)
Includes bibliographical references.
1. X-ray crystallography. I. Title.
QD945.T36   1976   548'.83   75-45196
ISBN 0 08 019692 6

SERIES EDITOR'S PREFACE

X-ray topography is the latest in a long line of tools for the study of crystals which use X-ray diffraction. This is the first book devoted entirely to developing the theme of the principles and application of this technique and will fill a very real gap in the literature. It is assured of a ready welcome. I am particularly glad to welcome it into this series as it compliments existing and forthcoming volumes, particularly Crystal Growth, Oxide Semiconductors, and Defects and Surfaces in Semiconducting Materials and Devices.

Many laboratories interested in the study of solid state devices and the materials which go into them now have X-ray topographic facilities. The technique has proved useful in the study of dislocations and faults in metal crystals, semiconducting materials and bubble memory garnets as well as a wide range of other materials from natural diamonds to silicon integrated circuits. The quality of heterojunction lasers was improved when the behaviour of dislocations in the (Ga,Al)As layers was illuminated by X-ray topographic studies. It is a very powerful new tool for solid state scientists and crystal growers and is already well established for the assessment of crystal quality in a wide range of single crystal materials.

Dr. Brian Tanner is an acknowledged leader in the field although still quite a young man. He studied at Oxford before going to Durham to set up a laboratory for X-ray topography in the Department of Physics. He has given invited papers at several International Conferences and published over twenty research papers.

Brian Pamplin
January, 1976.
Scientific Advisers and Co.,
15 Park Lane,
Bath.

PREFACE

Although by no means a new technique, the use of X-ray topography is at present increasing rapidly, particularly as an aid to crystal growth studies and quality control of monolithic crystal devices. The technique is complementary to its cousin, transmission electron microscopy in that X-ray topography enables a thick, nearly perfect single crystal to be examined with a relatively poor resolution over a large area whereas electron microscopy necessarily uses thin specimens of quite high dislocation density and examines a very small area with excellent resolution.

Improvements in crystal growth techniques in the last five years have provided many new materials suitable for X-ray topographic study and in turn, X-ray topography has provided the crystal grower with valuable data on the quality of his products. The feed-back between topographer and crystal grower or device manufacturer has proved so successful that today X-ray topographic analysis is performed as a standard routine by many crystal growing groups and firms manufacturing integrated circuits. The production of highly perfect single crystals has permitted the observation of many new X-ray optical phenomena and the detailed experimental verification of the various theories of dynamical X-ray diffraction. Correspondingly, the newly discovered effects have been utilized to the benefit of the crystal grower and device manufacturer in the development of new techniques with greater sensitivity to lattice parameter changes. As much of modern electronic engineering relies heavily on highly perfect single crystals, X-ray topography has claim to an important place in the hierarchy of assessment techniques.

Yet, despite the complex interaction between X-ray topographer and crystal grower, there seems to exist a certain lack of awareness of the recent developments, the potential and the problems of X-ray topography amongst non-specialists. There is a feeling that X-ray topography is a black art, understood only by a few initiates. While several excellent review articles have appeared, their existence is not widely known and in the sixteen years since X-ray topography was developed in its modern high resolution form, no book has been published on the subject.

From discussions with my colleagues, I am convinced that there exists a need for an elementary treatment of X-ray topography, comprehensible to the non-specialist who may have much to gain from occasional use of the techniques and it is this gap that the present volume is intended to fill. The book falls roughly into three sections. It is not comprehensive and my apologies are extended to any of my colleagues who may feel that their work has been unjustly neglected. In the first section the basic theory necessary for the understanding of topographic contrast is presented together with the chief experimental techniques and an analysis of the types of contrast observed. The second section presents some applications of topography. Those considered are included as being typical or classic studies and illustrate the kind of information obtained from topography. The final section reviews the work on assessment of crystal perfection in direct relation to the growth process. This is an area where future development will be rapid

and it is hoped that the growth points are anticipated. Following each
chapter, a selected bibliography of additional papers is included as an
appendix. These are grouped according to the main subject headings treated
in the text.

My thanks are extended to those of my colleagues throughout the world who
have provided the photographs without which the text would be lifeless.
In particular I would like to thank Dr. A.D. Milne for his thorough and
constructively critical reading of the manuscript and Mrs. S.M. Naylor for
typing the final copy of the text. I acknowledge with gratitude the support
of my wife and family, to whom the book is dedicated.

CONTENTS

List of Symbols Commonly Used in the Text   xiii

CHAPTER 1   Basic Dynamical X-ray Diffraction Theory   1

         1.1.   Fundamental equations of the dynamical theory
                in a perfect crystal   2

         1.2.   The dispersion surface   4

         1.3.   Anomalous transmission   6

         1.4.   Boundary conditions   8

         1.5   Energy flow   10

         1.6.   Pendellösung   14

         1.7.   Range of Bragg reflection   18

         1.8.   Generalized diffraction theory   20

         1.9.   Extension to asymmetric reflection   21

         1.10. Analysis   22

                References   23

CHAPTER 2   Experimental Techniques   24

         2.1.   Principles   24

         2.2.   The Berg-Barrett method   25

                2.2.1.   Reflection   26
                2.2.2.   Transmission (Barth-Hosemann geometry)   27

         2.3.   Lang's technique   28

         2.4.   Experimental procedures for taking Lang
                topographs   32

                2.4.1.   Setting up the crystal   32
                2.4.2.   Setting up the diffraction vector in
                         the horizontal plane   33

            2.4.3.   Finding the Bragg reflection   33
            2.4.4.   Recording the topograph   34

      2.5. Topographic resolution   35

            2.5.1.   Vertical resolution   35
            2.5.2.   Horizontal resolution   35

      2.6. Photography   38

      2.7. Enlargement of topographs   40

      2.8. Rapid high resolution topography   40

      2.9. Direct viewing of X-ray topographs   44

            2.9.1.   Direct conversion   44
            2.9.2.   X-ray to optical conversion   47

      2.10. Double crystal topography   47

      2.11. X-ray Moiré topography and interferometry   53

      2.12. Synchrotron topography   56

            References   59

            Appendix   61

CHAPTER 3   Contrast on X-ray Topographs   63

      3.1. Crystals without planar or line defects   63

            3.1.1.   Pendellösung fringes in traverse
                     topographs   64
            3.1.2.   Pendellösung fringes in section
                     topographs   67
            3.1.3.   Energy flow in section topographs   67

      3.2. Dynamical diffraction in distorted crystals   71

            3.2.1.   Small distortions   71
            3.2.2.   Large distortions   78

      3.3. Contrast of crystal defects in topographs   81

            3.3.1.   Dislocations in section topographs   81
            3.3.2.   Dislocations in traverse topographs   82
            3.3.3.   Contrast of precipitates   84
            3.3.4.   Surface damage   86
            3.3.5.   Contrast of stacking faults in section
                     topographs   86
            3.3.6.   Contrast or stacking faults in traverse
                     topographs   89

3.3.7.   Contrast of twins   93
3.3.8.   Contrast of magnetic domains   93
3.3.9.   Growth bands   95
3.3.10. Contrast on a non-ideal topograph   96

References   97
Appendix   98

CHAPTER 4   Analysis of Crystal Defects and Distortions   100

4.1.   Dislocations   100

4.1.1.   Analysis of Burgers vectors   100
4.1.2.   Study of the early stages of plastic
         deformation   102
4.1.3.   Studies of chemical attack   104
4.1.4.   Device control   104

4.2.   Planar defects   114

4.2.1.   Stacking faults   114
4.2.2.   Twins   116
4.2.3.   Ferroelectric domains   116
4.2.4.   Magnetic domains   119
4.2.5.   Ion implantation   122

References   124

Appendix   126

CHAPTER 5   Crystals Grown From Solution   130

5.1.   Growth from aqueous solution   130

5.1.1.   Dislocations in solution-grown crystals   134

5.2.   Hydrothermal growth   136

5.3.   Flux growth   139

References   143

Appendix   144

CHAPTER 6   Naturally Occurring Crystals   145

6.1.   Introduction   145

6.2.   Diamond   145

6.3.   Quartz   148

6.4.   Calcite, Magnesite, and Dolomite   150

6.5.   Fluorite   150

6.6.   Topaz and Apatite   151

6.7.   Barite, Mica, and Ice   152

6.8.   Résumé   152

        References   153

        Appendix   153

CHAPTER 7   Melt, Solid State, and Vapour Growth   155

7.1.   Melt growth   155

        7.1.1.   Semiconductors   155
        7.1.2.   Metals   158
        7.1.3.   Oxides   162
        7.1.4.   Ice   163

7.2.   Solid state growth   164

7.3.   Vapour growth   164

        7.3.1.   Whiskers   164
        7.3.2.   Metals   165
        7.3.3.   Inorganic crystals   166

        References   167

        Appendix   168

INDEX   171

LIST OF SYMBOLS COMMONLY USED
IN THE TEXT

A	Angular amplification
$\underline{b}$	Burgers vector
$\overline{\beta}$	Ray path parameter
c	Velocity of light
C	Polarization factor
c_{ij}	Elastic constants
γ_o, γ_g	Cosines of angle between Bragg planes and beam directions
$\underline{D}$	Electric displacement
$\overline{D}_o, D_g$	Component amplitudes of wavefields
d	Lattice plane spacing
δ, ε	Resolution
$\underline{E}$	Electric field
η	Deviation parameter
F_g	Structure factor
$\underline{g}, \underline{h}$	Reciprocal lattice vectors
$\overline{H}$	Magnetic field
$\underline{k}$	Wavevector *in vacuo*
$\underline{K}_o, \underline{K}_g$	Wavevector in the crystal
λ	Wavelength
Λ_o	Dispersion surface diameter
μ	Absorption coefficient
$\underline{P}$	Poynting vector
$\overline{p}$	Energy flow parameter
R	Amplitude ratio of diffracted and transmitted waves
r_e	Classical electron radius
$\hat{s}_o, \hat{s}_g$	Unit vectors in the transmitted and diffracted beam directions
χ_o	Susceptibility
χ_g	Fourier component of susceptibility
θ_B	Bragg angle
$\Delta\theta$	Angular deviation from exact Bragg angle
$\underline{u}$	Atomic displacement, dislocation line direction
$\overline{V}_c$	Volume of unit cell
ω	Angular frequency
ξ_g	Extinction distance

ACKNOWLEDGMENTS

The following photographs are copyright material and are reproduced here by kind permission of the publishers and authors.

Cover and Fig. 6.2	J. *Crystal Growth* (1974) <u>24/25</u>, 108.
Fig. 2.9(b,c)	J. *Crystal Growth* (1974) <u>24/25</u>, 64.
Fig. 2.16	*Nature* (1968) <u>220</u>, 652.
Fig. 3.3(a) and Fig. 7.2	J. *Appl. Phys.* (1973) <u>44</u>, 3905.
Fig. 3.12	J. *Appl. Cryst.* (1974) <u>7</u>, 372.
Fig. 4.1	J. *Mater. Sci.* (1972) <u>7</u>, 531.
Fig. 4.2	*Phil. Mag.* (1975) <u>32</u>, 283.
Fig. 4.4	J. *Appl. Cryst.* (1973) <u>6</u>, 31.
Fig. 4.9	J. *Appl. Phys.* (1967) <u>38</u>, 3495.
Fig. 4.10	J. *Crystal Growth* (1974) <u>24/25</u>, 637.
Fig. 4.11	*Phil. Mag.* (1973) <u>28</u>, 1015.
Fig. 5.3	J. *Crystal Growth* (1973) <u>18</u>, 135.
Fig. 5.4	*Acta Cryst.* (1973) <u>A29</u>, 495.
Fig. 5.5	J. *Crystal Growth* (1974) <u>24/25</u>, 541.
Fig. 5.7	J. *Crystal Growth* (1975) <u>29</u>, 281.
Fig. 5.8	J. *Crystal Growth* (1975) <u>28</u>, 77.
Fig. 6.1	*Appl. Phys. Lett.* (1969), <u>15</u>, 258.
Fig. 7.4(a)	J. *Crystal Growth* (1974), <u>24/25</u>, 527.

CHAPTER 1

BASIC DYNAMICAL X-RAY DIFFRACTION THEORY

It is important from the outset to emphasize that X-ray topography is not concerned primarily with the study of surfaces. The full title, X-ray Diffraction Topography, is much clearer as it indicates that the topography we are studying is that of the diffracting planes in the crystal, not the topography of the exterior features. Of course, the contours of the crystal surfaces are important in determining the contrast on X-ray topographs, but this is of somewhat secondary importance to the contours of the crystal lattice planes. When we use the technique to observe dislocations, we are studying the topography of the lattice planes around the defect. We do this by recording the intensity of the X-rays diffracted from the deformed planes which differs from the intensity diffracted by the perfect crystal forming a localized image of the defect. Essentially, we are using the phenomenon of diffraction to probe the internal structure of the crystal. It is not, however, a point probe, and the interpretation of the observed contrast is far from trivial.

At the simplest level, we can obtain some insight into how dislocations are imaged in the following way. Consider a perfect crystal set to diffract monochromatic X-radiation of wavelength λ from a set of lattice planes spaced d. For a strong diffracted beam to emerge at angle $2\theta_B$ to the incident beam the well known Bragg relation applies. That is,

$$\lambda = 2d \sin \theta_B \qquad (1.1)$$

It is clear that when the lattice spacing or lattice plane orientation varies locally, e.g. around a dislocation, the relation will not apply simultaneously to the perfect and distorted regions. Consequently there is a difference in intensity corresponding to the two regions, i.e. an image of the defect.

In order to interpret these changes in intensity, and more importantly, relate them to the lattice plane topography in a particular crystal under investigation, we need to know something about the theory of X-ray diffraction in solids. Now although an elementary treatment of X-ray diffraction may be found in many textbooks on solid state physics, the treatment is based on the KINEMATICAL approximation. In this situation, it is assumed that the amplitudes of the scattered waves are at all times small compared with the incident wave amplitude. For small crystals, of dimensions less than about a micrometre in diameter, and in heavily deformed crystals where the dislocations act to divide the crystal into a mosaic structure of independently diffracting cells, the kinematical theory may be employed satisfactorily to obtain information on the crystal structure. However, for large single crystals which are also highly perfect, the amplitude of a diffracted wave becomes comparable with that of the incident beam. Interchange of energy

1

occurs between the beams as they pass through the crystal and a kinematical theory containing an extinction correction cannot be applied. It is necessary to develop a DYNAMICAL theory of diffraction to account for the processes occurring.

The problem can be treated in a variety of ways, and the theory of Darwin, in which the scattered amplitude due to an elementary layer of material is used to obtain a set of differential equations, was originally applied to the two-beam theory of electron diffraction (see Whelan, 1970). However, a more satisfactory approach is that of von Laue (1952) and this is followed here. The dynamical theory in the X-ray case has been excellently reviewed by Batterman and Cole (1964) and Authier (1970) and the reader is strongly recommended to read these articles. More general treatments are to be found in the books by James (1948), Zacharaisen (1945) and von Laue (1960). The article by Hart (1971) on Bragg reflection X-ray optics, referred to again in Chapter 2, also contains a concise summary of the elements of the theory.

1.1.   FUNDAMENTAL EQUATIONS OF THE DYNAMICAL THEORY IN A PERFECT CRYSTAL

The problem can be stated with deceptive simplicity. We require a solution of Maxwell's equations in a periodic medium matched to solutions which are plane waves outside the crystal.

The solutions obtained must reflect the periodicity of the crystal lattice, and such functions are known as Bloch functions. The Bloch waves can be constructed from a superposition of plane waves, but it has been verified experimentally that the Bloch waves do have physical significance and are not merely convenient mathematical constructions.

According to the kinematical theory of X-ray diffraction, each diffracted wave is associated with a vector in reciprocal space corresponding to a reciprocal lattice point $\underline{g}$ and also has a wavevector inside the crystal $\underline{K}_g$.

The diffracted beam wavevector is related to the wavevector of the incident wave in the crystal by the Laue equation,

$$\underline{K}_g = \underline{K}_o + \underline{g}. \tag{1.2}$$

In the dynamical case, which must yield the same results as the kinematical theory in the limit of thin crystals, we may expect solutions to consist of linear combinations of such waves. Accordingly, we look for solutions of Maxwell's equations for the electric displacement $\underline{D}$ of the form

$$\underline{D} = \sum_g \underline{D}_g \exp(-2\pi i \underline{K}_g \cdot \underline{r}) \exp(i\omega t). \tag{1.3}$$

Assuming that the electrical conductivity is zero and the magnetic permeability is unity, Maxwell's equations reduce to

$$\text{curl curl } \underline{D} = - \{(1 + \chi)/c^2\}(\partial^2 \underline{D}/\partial t^2) . \tag{1.4}$$

In a periodic medium, the susceptibility χ is periodic and can be expanded as a Fourier series over the reciprocal lattice as

$$\chi = \sum_h \chi_h \exp(-2\pi i \underline{h} \cdot \underline{r}). \tag{1.5}$$

χ_h is related to the structure factor F_h by a proportionality constant

$$\chi_h = -r_e \lambda^2 F_h / \pi V_c,$$ (1.6)

where r_e is the classical electron radius and V_c is the volume of the unit cell. The structure factor is, of course, related to the atomic scattering factors by

$$F_h = \underset{\substack{\text{unit}\\ \text{cell}}}{\Sigma}\ f_i\ \exp(2\pi i \underline{h}.\underline{r}_i),$$ (1.7)

where $\underline{r}_i$ is the position vector in real space of the i th atom with respect to the origin.

As χ is very small in the X-ray region, typically of the order of 10^{-5}, we can therefore write (1.4) as

$$\text{curl curl } (1 - \chi)\underline{D} = -(1/c^2)\partial^2 \underline{D}/\partial t^2$$ (1.8)

and substitution of (1.3) and (1.5) into (1.8) leads, after some manipulation, to

$$\underset{h}{\Sigma}\{\chi_{g-h}(\underline{K}_g.\underline{D}_h)\underline{K}_g - \chi_{g-h}(\underline{K}_g.\underline{K}_g)\underline{D}_h\} = \{k^2 - (\underline{K}_g.\underline{K}_g)\}\underline{D}_g,$$ (1.9)

where $k = \omega/c$ is the vacuum wavevector.

These equations are the fundamental equations of the dynamical theory and are a vector form of the equivalent equations which may be obtained in the electron case by solving Schrodinger's equation in a periodic medium. Unlike the electron case, it is fortunate in X-ray diffraction that only very rarely does more than one reciprocal lattice point provide a diffracted wave of appreciable amplitude. This arises from the much larger radius of curvature of the Ewald sphere in the electron case compared with X-ray diffraction. Thus, we need only consider two waves to have appreciable amplitude in the crystal - that associated with the incident wave and that associated with the diffracted wave from a reciprocal lattice vector $\underline{g}$. Equations (1.9) then reduce to

$$\chi_g(\underline{K}_g.\underline{D}_o)\underline{K}_g - \chi_g(\underline{K}_g.\underline{K}_g)\underline{D}_o + \chi_o(\underline{K}_g.\underline{D}_g)\underline{K}_g - \chi_o(\underline{K}_g.\underline{K}_g)\underline{D}_g$$
$$= (k^2 - \underline{K}_g.\underline{K}_g)\underline{D}_g$$ (1.10a)

$$\chi_{\bar{g}}(\underline{K}_o.\underline{D}_g)\underline{K}_o - \chi_{\bar{g}}(\underline{K}_o.\underline{K}_o)\underline{D}_g + \chi_o(\underline{K}_o.\underline{D}_o)\underline{K}_o - \chi_o(\underline{K}_o.\underline{K}_o)\underline{D}_o$$
$$= (k^2 - \underline{K}_o.\underline{K}_o)\underline{D}_o.$$ (1.10b)

Taking the scalar product of (1.10a) with $\underline{D}_g$ and (1.10b) with $\underline{D}_o$ and remembering that waves of electric displacement are always transverse (i.e. $\underline{K}_o.\underline{D}_o = \underline{K}_g.\underline{D}_g = 0$), we obtain

$$k^2 C\chi_{\bar{g}} D_g + \{k^2(1 + \chi_o) - \underline{K}_o \cdot \underline{K}_o\} D_o = 0,$$

$$\{k^2(1 + \chi_o) - \underline{K}_g \cdot \underline{K}_g\} D_g + k^2 C\chi_g D_o = 0,$$

(1.11)

where $\quad C = \underline{D}_o \cdot \underline{D}_g = 1$ for σ polarization

$\qquad\qquad\qquad = \cos 2\theta_B$ for π polarization.

For a non-trivial solution

$$\begin{vmatrix} k^2 C\chi_{\bar{g}} & k^2(1 + \chi_o) - \underline{K}_o \cdot \underline{K}_o \\ k^2(1 + \chi_o) - \underline{K}_g \cdot \underline{K}_g & k^2 C\chi_g \end{vmatrix} = 0. \quad (1.12)$$

Writing

$$\alpha_o = \tfrac{1}{2}k\{\underline{K}_o \cdot \underline{K}_o - k^2(1 + \chi_o)\},$$

$$\alpha_g = \tfrac{1}{2}k\{\underline{K}_g \cdot \underline{K}_g - k^2(1 + \chi_o)\},$$

(1.13)

We arrive at

$$\alpha_o \alpha_g = \tfrac{1}{4}k^2 C^2 \chi_g \chi_{\bar{g}}. \quad (1.14)$$

1.2. THE DISPERSION SURFACE

We are now in a position to introduce one of the most important and useful concepts relating to dynamical diffraction theory - the dispersion surface. Equation (1.14) is the fundamental relation linking $\underline{K}_o$ and $\underline{K}_g$ in the crystal, and we can represent it geometrically in the following way. About the origin O and the reciprocal lattice point G (where $\vec{OG} = \underline{g}$) draw spheres of radius k. A section through these spheres is shown in Fig. 1.1. Only close to the intersection, the Laue point, will strong diffraction occur as only there is the Laue equation (1.2) satisfied. Now it is easy to see from (1.11) that if no diffracted wave exists (i.e. $\underline{D}_g = 0$), then

$$|\underline{K}_o| \simeq k(1 + \chi_o/2). \quad (1.15)$$

That implies that the wavevector of the wave in the crystal is given by the wavevector in vacuo multiplied by the refractive index. As χ_o is small, $|\underline{K}_o| \simeq k$. A second pair of spheres are drawn about O and G with radius $k(1 + \chi_o/2)$. Far from the Laue point, the tail of the wavevector in the crystal $\underline{K}_o$ will fall on the sphere about O. However, when strong diffraction occurs (1.14) defines the relation between $\underline{K}_o$ and $\underline{K}_g$ and thus the tail cannot lie on the spheres. Figure 1.2 shows the region close to

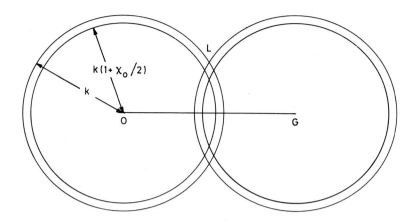

Fig. 1.1. Spheres in reciprocal space about the lattice points
          O and G showing the position of the Laue point L.

the Laue point at a very much greater magnification.

The arcs AB and A´B´ correspond to the spheres of radius k and the arcs
CD and C´D´ to those of radius $k(1 + \chi_o/2)$. The tails of the wavevectors $\underline{K}_o$
and $\underline{K}_g$ lie on the solid curves. We note that α_o and α_g correspond to the
perpendicular distances from the point P to the spheres CD and C´D´. As the
region is very small compared to the radius of the spheres, the spheres may
be approximated as planes. Then the equation of the dispersion surface
(1.4) becomes a hyperboloid of revolution with axis OG. Our section of the
dispersion surface is thus a hyperbola asymptotic to CD and C´D´. There are
four branches - two for each polarization state, the upper ones being
denoted branch 1 and the lower ones branch 2. For small Bragg angles the
dispersion surfaces of the σ and π polarizations are very close together,
but at higher Bragg angles the effects of polarization become extremely
important.

Any wave propagating in the crystal must have wavevectors $\underline{K}_o$ and $\underline{K}_g$ lying
on the dispersion surface, and returning to (1.3) we see that the wave has
amplitude

$$\underline{D} = \exp(i\omega t)\{\underline{D}_o \exp(-2\pi i\underline{K}_o \cdot \underline{r}) + \underline{D}_g \exp(-2\pi i\underline{K}_g \cdot \underline{r})\}, \qquad (1.16)$$

where the amplitudes $\underline{D}_o$ and $\underline{D}_g$ are determined from (1.11) and (1.14).
The amplitude ratio

$$R = D_g/D_o = 2\alpha_o/Ck\chi_{\bar{g}} = Ck\chi_g/2\alpha_g. \qquad (1.17)$$

Thus not only can we determine the wavevector from the position of the tie
points P on the dispersion surface but also the amplitude. The importance
of the construction becomes apparent.

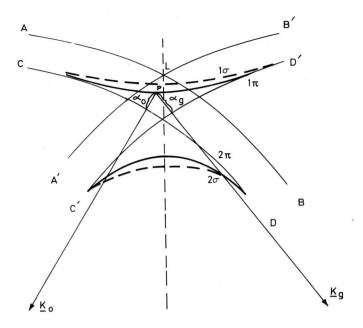

Fig. 1.2. The dispersion surface construction. This is a magnified
         drawing of the region about L. The degeneracy is lifted
         by the crystal potential and the tails of the allowed
         wavevectors lie on the bolder-drawn curves.

1.3. ANOMALOUS TRANSMISSION

Utilizing (1.2), we note that (1.16) becomes, neglecting time dependence,

$$\underline{D} = \{\underline{D}_o + \underline{D}_g \exp(-2\pi i \underline{g}\cdot\underline{r})\}\exp(-2\pi i\underline{K}_o\cdot\underline{r}).\tag{1.18}$$

The wave in the crystal is not a plane wave, it is a Bloch wave and the
amplitude is modulated with a periodicity corresponding to the Bragg planes.
In terms of intensity

$$I = D^2 = D_o^2\{1 + R^2 + 2RC\cos(2\pi\underline{g}\cdot\underline{r})\}.\tag{1.19}$$

The intensity is modulated by the factor $\cos(2\pi g.r)$ which has maxima at
$\underline{g}\cdot\underline{r} = n$ and minima at $\underline{g}\cdot\underline{r} = (2n + 1)/2$ with n integral. That is, the maxima
and minima of the standing wavefield occur either at or halfway between
the atomic planes, as $\underline{g}\cdot\underline{r} = n$ corresponds to a plane of atoms. This is
illustrated schematically in Fig. 1.3.
   Now two things may be noted. Firstly, that the sign of R determines
whether maxima and minima occur at the atomic planes. We can see from
(1.13) and (1.17) that the sign of R is opposite for Bloch waves with tie
points on opposite branches of the dispersion surface. Thus for one Bloch

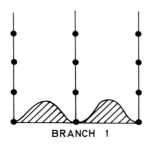
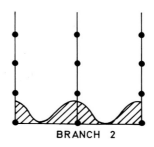

BRANCH 1   BRANCH 2

Fig. 1.3. Intensity of the two wavefields at the exact Bragg
          condition in the 200 reflection. Note that the branch 1
          waves have a minimum intensity at the atomic positions.

wave, the intensity maxima occur at the atomic planes, for the other the
minima occur. The wave with R negative has minima at the atomic planes.
Secondly, the amplitude of the modulations is a function both of R and C.
Maximum modulation occurs when both R and C equal unity, that is in a
centrosymmetric crystal, for the σ polarized wave at the exact Bragg
condition.

We might have anticipated this spatial modulation earlier, as one sees
from (1.5) that the Fourier components of susceptibility χ_h remain
unchanged if the origin is displaced by a multiple of the Bragg plane
spacing.

Thus far, no mention has been made of absorption processes. It is
possible to take absorption into account phenomenologically by making the
structure factor, and hence the wavevectors, complex. The magnitude of the
imaginary part then corresponds to the absorption coefficient. This
approach will not be adopted here; for details the reader is referred to
one of the review articles quoted earlier. Without further mathematical
derivation, however, we can obtain an important qualitative insight into the
behaviour of X-rays in thick, nearly perfect, crystals.

Absorption occurs mainly by the photoelectric process at crystallographic X-ray wavelengths, and as the electron density is greatest at the
atomic planes, the Bloch wave with intensity maxima at the atomic planes
will have the greater photoelectric absorption cross-section. For each
polarization state, the waves with R negative are less strongly absorbed
than those with R positive. For example, in the 200 reflection in NaCℓ, the
structure factor F_g is positive. Hence, χ_g is negative and the branch 2
wavefield is absorbed more strongly than branch 1. This effect is known as
anomalous transmission or the Borrmann effect and was discovered as early as
1941 by Borrmann in calcite. Maximum modulation of the branch wavefield in
NaCℓ, and thus minimum absorption of that wavefield, occurs for the 1σ
polarized wave at the exact Bragg condition. Thus, in a very thick crystal
we find that only 1σ polarized waves at the exact Bragg angle are transmitted. The crystal acts as a polarizer.

Under suitable conditions, the anomalous transmission effect can be very
pronounced and it is therefore possible to obtain a transmitted wave through
crystals whose product of average linear absorption coefficient and thickness is over 30. (We would naively expect the wave to be attenuated by a
factor of exp (−30) in this case!) Anomalous transmission is a feature of a

perfectly periodic lattice, and imperfections of any kind tend to reduce or destroy it. Dislocations, for example, totally destroy the effect, and this was one of the earliest methods by which dislocations were observed directly (Borrmann, Hartwig, and Irmler, 1958).

Point defects, which have a shorter range strain field significantly reduce the amount of anomalous transmission and Dederichs (1970) has developed a theory relating the transmission to the defect concentration and size. It has been applied to silicon (Patel, 1973) and aluminium (Nøst, Larson, and Young, 1972) with some considerable success. The loop sizes deduced using Dederichs' theory from anomalous transmission and small angle scattering measurements give reasonable agreement with the sizes measured directly by transmission electron microscopy.

Anomalous transmission decreases as the temperature rises and best results are obtained at cryogenic temperatures. As the atoms are not static, the thermal vibration spreads the electron charge over a wider range than on our simplistic diagram of Fig. 1.3. The mean square amplitude of vibration increases as the temperature rises, thus reducing the amount of transmission. Not surprisingly, we find the Debye-Waller factor appearing in the expression for the absorption coefficient. We also note that even at very low temperature, the anomalous transmission will never be perfect, for although the electric field is zero at the atom centres, it is not zero at the K and L shells where most of the photoelectric absorption takes place. Some absorption will always occur.

In the example quoted previously, that of NaCl 200 reflection, it was the branch 1 wavefield which was best transmitted. It turns out that in the diamond and rocksalt structures, the intensity minima lie always at the atomic planes. However, by use of suitable absorption edges it is possible to find crystals in which the planes of principal absorbing atoms lie between the principal scattering planes. Then the branch 2 wavefield is best transmitted, and this effect has been demonstrated by Meieran and Blech (1968).

Derivation of the absorption coefficient μ of the two Bloch waves is tedious even for waves at the centre of the dispersion surface. We simply quote the result, that is,

$$\mu(\text{zero}) = \mu_o \left(1 \pm |C| \left|\frac{\chi_{ig}}{\chi_{io}}\right| \exp\text{-}M \cos \phi\right), \tag{1.20}$$

where μ_o is the average absorption coefficient, M is the Debye-Waller factor, χ_{ig} and χ_{io} are the imaginary parts of the Fourier coefficients of the susceptibility, and ϕ is the phase angle between the hkl coefficients of the real part of the susceptibility. We find contained in this formula all the features just discussed from a less mathematical standpoint.

1.4. BOUNDARY CONDITIONS

The fundamental equations determine the propagation inside the crystal, but it is vital to our experiments that we are able to determine the wave amplitudes outside the crystal as it is outside that all measurements are made. As in any wave propagation problem we must ensure that no discontinuities occur at the boundary. The matching conditions for electromagnetic waves are well known, and imply constraints on frequency, amplitude, and

wavevector. The requirement of continuity of frequency does not explicitly concern us, but the continuity of wavevector has important consequences.
Consider a plane polarized plane wave of amplitude

$$\underline{D} = \underline{D}_i \, \exp(-2\pi i \underline{k}_e \cdot \underline{r}) \tag{1.21}$$

incident on the upper surface of a crystal. (We can generalize to the unpolarized state by superposition of waves, and although we must assume here that the surface is smooth on the scale of the X-ray wavelength (!) there are more aesthetically pleasing ways of arriving at the same result.) Any Bloch wave in the crystal can be considered as a superposition of plane waves

$$\underline{D} = \sum_g \underline{D}_g \, \exp(-2\pi i \underline{K}_g \cdot \underline{r}) \tag{1.22}$$

and the conditions that the wavevector be continuous across the boundary requires

$$\exp(-2\pi i \underline{k}_e \cdot \underline{\tau}) = \exp(-2\pi i \underline{K}_g \cdot \underline{\tau}) \tag{1.23}$$

for each plane wave. $\underline{\tau}$ is any unit vector in the surface. This is equivalent to saying that the tangential component of wavevector must be equal for plane waves on either side of the boundary. Thus wavevectors inside the crystal differ from that outside only by a vector normal to the crystal surface, i.e.

$$\underline{K}_{o_i} - \underline{k}_e = \underline{K}_{g_i} - \underline{k}_e = \delta\hat{n}, \tag{1.24}$$

where $\hat{n}$ is a unit vector normal to the surface and $i = 1$ or 2 depending on the branch of the dispersion surface.
This relation becomes more instructive if depicted graphically as in Fig. 1.4. There we see that the tie points excited inside the crystal by an external wave can be found simply by drawing a vector normal to the crystal surface from the tip of the wavevector $\underline{k}_e$ of the incident wave. The tie points P and Q are given directly by the points of intersection of this vector $\hat{n}$ with the dispersion surface. A plane wave thus excites two tie points. In transmission (the Laue geometry) these lie on opposite branches of the dispersion surface (Fig. 1.4a). In reflection, the Bragg geometry, the tie points lie on the same branch of the dispersion surface (Fig. 1.4b). It is interesting to note that in the Bragg case there exists a region in which no tie points are excited and waves cannot propagate into the crystal. These evanescent waves are analogous to the evanescent waves in total internal reflection of light in which the wave amplitude decays exponentially on proceeding into the second medium.
The third condition is on the field amplitudes. In the X-ray region, the refractive index is very close to unity and to a good approximation, refraction at the boundary can be ignored. The electric displacement vectors can then be equated on either side of the boundary. Thus if the electric displacement amplitude outside the crystal is $\underline{D}_i$ we can write

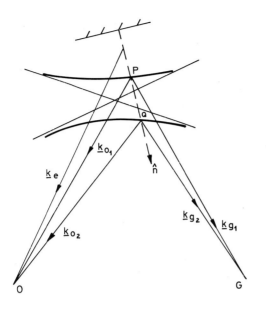

$$\left. \begin{array}{l} \underline{D}_i = \underline{D}_{o_1} + \underline{D}_{o_2}, \\[6pt] 0 = \underline{D}_{g_1} + \underline{D}_{g_2} \end{array} \right\} \qquad (1.25)$$

In some texts the electric field vectors are used and longitudinal parts neglected. This is, however, equivalent to working with the (transverse) electric displacement and then neglecting refraction at the boundary.

   In the following sections we will determine some of the properties of the waves in the crystal. For simplicity we will confine discussion to the case of symmetric diffraction. Extension to the asymmetric case is straightforward and can be found in the review articles referenced. The corresponding formulae for asymmetric reflection corresponding to the more important of the results will be given at the end of the chapter.

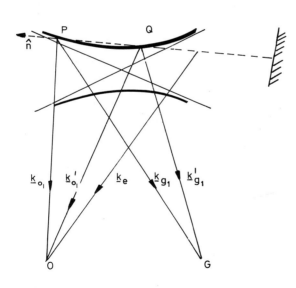

Fig. 1.4. Geometrical representation of the boundary conditions.
   (a) Laue, transmission geometry.
   (b) Bragg, reflection geometry.

1.5.   ENERGY FLOW

   Consider a crystal set in the symmetric Laue geometry as shown in Fig. 1.5. The direction of energy flow is given by the Poynting vector

$$\underline{P} = \underline{E} \times \underline{H} \qquad (1.26)$$

and this, when averaged over the unit cell and over time can be written

$$\underline{P} = I_o \hat{\underline{s}}_o + I_g \hat{\underline{s}}_g, \qquad (1.27)$$

where I_o and I_g are the intensities of the refracted and diffracted beams respectively and $\hat{\underline{s}}_o$ and $\hat{\underline{s}}_g$ are unit vectors parallel to

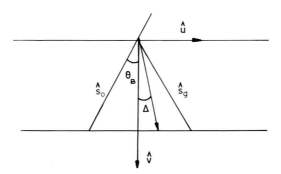

Fig. 1.5. The Borrmann fan showing the direction of energy flow
         with respect to the coordinate system used in the text.

the incident and reflected directions.
   With $\hat{u}$ and $\hat{v}$ unit vectors perpendicular and parallel to the reflecting
planes we have for the direction of propagation of any wavefield

$$\left. \begin{array}{l} \underline{P}.\hat{u} = P \sin \Delta = I_o \hat{s}_o.\hat{u} + I_g \hat{s}_g.\hat{u} = (I_g - I_o)\sin \theta_B, \\[2mm] \underline{P}.\hat{v} = P \cos \Delta = I_o \hat{s}_o.\hat{v} + I_g \hat{s}_g.\hat{v} = (I_g + I_o)\cos \theta_B. \end{array} \right\} \qquad (1.28)$$

Here, Δ is the angle made between the propagation direction of the wavefield
and the Bragg planes. This is clearly given by

$$\tan \Delta = \tan \theta_B \frac{(I_g/I_o - 1)}{(I_g/I_o + 1)} = \tan \theta_B \frac{(R^2 - 1)}{(R^2 + 1)}. \qquad (1.29)$$

We now have to relate the intensities of the components of any arbitrary
wavefield to the position on the dispersion surface.
   In the coordinate system Ox, Oz in Fig. 1.6, the equation of the dispersion surface is

$$z^2 = \tfrac{1}{4}\Lambda_o^2 + x^2 \tan^2 \theta_B, \qquad (1.30)$$

where Λ_o is the diameter of the dispersion surface. The angle Θ between the
Oz axis and the normal to the dispersion surface is therefore given by

$$\tan \Theta = dz/dx = x \tan^2 \theta_B/z. \qquad (1.31)$$

We now introduce an extremely important parameter η which is called the
deviation parameter and determines how far from the exact Bragg position the
tie points are. In the symmetric situation η is defined as

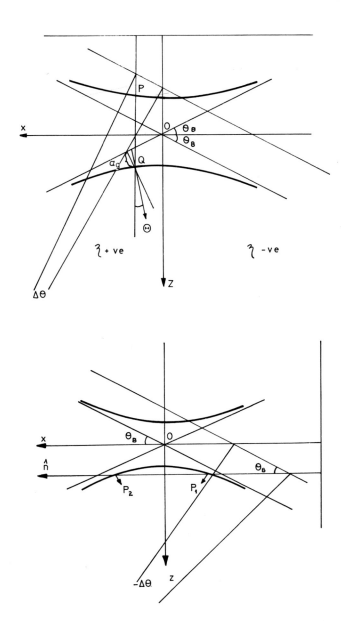

Fig. 1.6. (a) Dispersion surface construction giving the coordinate system used in the text for the Laue geometry. Tie points P and Q are excited by an incident wave at an angle Δθ from the exact Bragg condition. (b) Dispersion surface construction for the Bragg case.

$$\eta = 2x \tan \theta_B/\Lambda_o. \tag{1.32}$$

The angular deviation from the exact Bragg angle $\Delta\theta$ is given by

$$\Delta\theta = x \sec \theta_B/k, \tag{1.33}$$

where $\Delta\theta$ is negative when the angle of incidence is less than the Bragg angle. We then see that

$$\eta = 2 \sin \theta_B \; k\Delta\theta/\Lambda_o. \tag{1.34}$$

The deviation parameter is thus proportional to the angular deviation from the Bragg condition. In terms of this parameter, (1.30) becomes

$$z = \pm\Lambda_o(1 + \eta^2)^{\frac{1}{2}}/2. \tag{1.35}$$

Now the perpendicular distance from the tie point to the sphere about G, which is α_g, is given by

$$\alpha_g \sec \theta_B = z - x \tan \theta_B, \tag{1.36}$$

and thus (1.30) can also be written

$$(\alpha_g \sec \theta_B)^2 + \Lambda_o\eta\alpha_g \sec \theta_B - \tfrac{1}{4}\Lambda_o^2 = 0 \tag{1.37}$$

using (1.32) and (1.36).
Hence

$$\alpha_g \sec \theta_B = \tfrac{1}{2}\Lambda_o\{-\eta \pm (1 + \eta^2)^{\frac{1}{2}}\}. \tag{1.38}$$

In terms of the amplitude ratio $R = D_g/D_o$, we find

$$R = \eta \pm (1 + \eta^2)^{\frac{1}{2}}. \tag{1.39}$$

where the upper sign corresponds to branch 1 and the lower to branch 2. In deriving (1.39) we have used (1.17) and noted that at the centre of the dispersion surface, the diameter

$$\Lambda_o = 2\alpha_o(0) \sec \theta_B = 2\alpha_g(0) \sec \theta_B. \tag{1.40}$$

Hence

$$\Lambda_o^2 = \sec^2 \theta_B \; k^2 C^2 \chi_g \chi_{\bar{g}}. \tag{1.41}$$

Substitution of (1.39) into (1.31) using (1.32) and (1.35) yields

$$\tan \Theta = \tan \theta_B \; \frac{(R^2 - 1)}{(R^2 + 1)} \qquad\qquad (1.42)$$

Thus

$$\Delta = \Theta \qquad\qquad (1.43)$$

This is of great importance. We see that the direction of propagation is in a direction normal to the dispersion surface. Thus for any Bloch wave whose position of tie point we can determine from the boundary conditions we can immediately determine the direction of propagation of energy, i.e. the ray direction. At the exact Bragg condition where the tie points lie on the Brillouin zone boundary, the energy flow is parallel to the lattice planes. Thus in a thick crystal, where all wavefields except those at the Bragg condition (which undergo appreciable anomalous transmission) are absorbed, the energy flow is entirely parallel to the lattice planes.

Away from the exact Bragg condition the direction of energy flow of wavefields associated with the points excited on the two branches is not the same. This is of great importance. Consider an incident plane wave exciting two tie points P and Q (Fig. 1.6). If the spatial extent of the wave is limited, then in thick crystals the two waves become spatially separated. This was first demonstrated by Authier (1960), and subsequent experiments by his group have confirmed that the concept of a Bloch wave is not a mathematical fiction, but does correspond to the real situation inside a crystal. For example, once the waves have been separated, they act independently. When a dislocation line is cut by one Bloch wavefield and not the other, an image appears in the diffracted beam corresponding to one Bloch wave only, (Authier, Balibar, and Epelboin, 1970). The experiment is sketched in Fig. 1.7. A thick crystal is used to select a wave of limited spatial extend. If this is selected from the centre of the Borrmann fan, it can be shown that this wave has a very narrow angular divergence. The wave incident on the second crystal is thus a pseudo-plane wave and excites only two tie points. Due to the small width of the beam, the wavefields separate in the second crystal and in the example just quoted, the image of the dislocation line D is seen at A but not B.

Further verification of the physical reality of the wavefields has come from experiments by Hart and Milne (1971). In a similar experiment to Authier's (1961), but this time fabricating the two crystals from one monolithic block of silicon, they were able to show that the wavefield, and hence the ray, concept was retained in slightly distorted crystals.

1.6.   PENDELLÖSUNG

It is by now clear that an incident plane wave excites two Bloch waves and that the wavevectors associated with these tie points differ. The difference in wavevector leads to a difference in the propagation velocity, and hence we may expect interference effects to occur between the Bloch waves. A similar situation occurs in a birefringent crystal where the two normal modes of propagation are linearly polarized plane waves travelling with different velocities. Interference effects are observed in wedge shaped crystals (see Fathers and Tanner, 1973, for examples), a fringe occurring wherever the phase difference becomes a multiple of 2π.

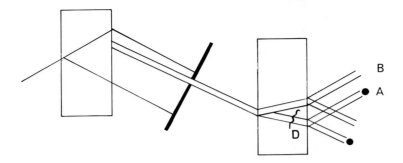

Fig. 1.7. Schematic diagram of the double crystal spectrometer used by
Authier and coworkers to obtain a pseudo-plane wave in the
second crystal. Here, due to the spatial separation of the
Bloch wavefields the image of the dislocation appears in
only one of the diffracted and transmitted images.

Fringes are observed in X-ray topographs of wedge-shaped crystals due to
the interference effects. They do not occur when only one Bloch wave is
present, e.g. in thick crystals in which anomalous absorption has left only
the best transmitted wave with appreciable amplitude. In an experiment
described in more detail later, Malgrange and Authier (1965) have shown that
when the wavefields are separated in space, fringes are also not observed.

Let us consider the specific case of the symmetric Laue geometry with an
incident plane wave of unit amplitude. Then the boundary conditions given
in (1.25) become

$$\left.\begin{array}{l} 1 = D_{o_1} + D_{o_2}, \\[1em] 0 = D_{g_1} + D_{g_2}. \end{array}\right\} \tag{1.44}$$

The amplitude ratio D_{g_i}/D_{o_i} for each wave is determined by the position of
the tie point. From (1.38) we have

$$R_i = D_{g_i}/D_{o_i} = \eta \pm (1 + \eta^2)^{\frac{1}{2}}, \tag{1.45}$$

where the positive sign refers to branch 1 and the negative sign to branch 2.
Introducing a new parameter β given by

$$\eta = \cot \beta, \tag{1.46}$$

we find from (1.44) and (1.45)

$$D_{o_1} = \cos^2 \beta/2, \qquad\qquad D_{o_2} = \sin^2 \beta/2,$$

$$D_{g_1} = -\sin \beta/2 \cos \beta/2, \qquad D_{g_2} = \sin \beta/2, \cos \beta/2. \qquad (1.47)$$

The amplitudes of the beams emerging from the crystal at depth t are found by applying the boundary conditions (1.25) again.

$$D_o^e = D_{o_1}(t) + D_{o_2}(t),$$

$$D_g^e = D_{g_1}(t) + D_{g_2}(t), \qquad\qquad (1.48)$$

where

$$D_{o_i}(t) = D_{o_i} \exp(-2\pi i t \underline{K}_{o_i} \cdot \hat{n}),$$

$$D_{g_i}(t) = D_{g_i} \exp(-2\pi i t \underline{K}_{g_i} \cdot \hat{n}), \qquad\qquad (1.49)$$

with $\hat{n}$ a unit vector normal to the surface and i = 1 or 2.
For the diffracted beam this yields an intensity I_g, where

$$I_g = D_g^{e*} D_g^e = D_{g_1}^2 + D_{g_2}^2 + 2D_{g_1} D_{g_2} \cos 2\pi t (\underline{K}_{g_2} - \underline{K}_{g_1}) \cdot \hat{n}$$

$$= \tfrac{1}{2} \sin^2 \beta \{1 - \cos 2\pi t (\underline{K}_{g_2} - \underline{K}_{g_1}) \cdot \hat{n}\}. \qquad (1.50)$$

Now from Figs. 1.4 and 1.6 we see

$$(\underline{K}_{g_1} - \underline{K}_{g_2}) \cdot \hat{n} = 2z = \Lambda_o (1 + \eta^2)^{\frac{1}{2}}, \qquad\qquad (1.51)$$

and thus in terms of η

$$I_g = \sin^2 (\pi \Lambda_o t (1 + \eta^2)^{\frac{1}{2}} / (1 + \eta^2). \qquad\qquad (1.52)$$

Similarly, I_o, the intensity of the transmitted beam is

$$I_o = \cos^2 (\pi \Lambda_o t (1 + \eta^2)^{\frac{1}{2}} / (1 + \eta^2). \qquad\qquad (1.53)$$

Equations (1.52) and (1.53) show the interference referred to above most clearly. The intensity of both diffracted and transmitted beams exhibits a periodic variation with crystal thickness. The period is the same for both beams and is given by a depth $\{\Lambda_o \, (1 + \eta^2)^{\frac{1}{2}}\}^{-1}$. This has a maximum value at $\eta = 0$, that is, at the exact Bragg condition. The depth corresponding to one period here is known as the extinction distance and is given by the reciprocal of the dispersion surface diameter. That is

$$\xi_g = \Lambda_o^{-1} = \cos \theta_B / Ck (\chi_g \chi_{\bar{g}})^{\frac{1}{2}} = \pi V_c \cos \theta_B / r_e \lambda C (F_g F_{\bar{g}})^{\frac{1}{2}}. \qquad (1.54)$$

For values of $\eta \neq 0$, the effective extinction distance is

$$\xi'_g = \xi_g / (1 + \eta^2)^{\frac{1}{2}}. \qquad (1.55)$$

As $\eta \rightarrow \infty$ the effective extinction distance decreases to zero as only one wave is excited in the crystal and no interference occurs.

We note further that the intensity of the diffracted and transmitted beams is complementary. Effectively, energy is being passed from one beam to the other as the wave passes through the crystal, and this prompted Ewald to name the phenomenon "Pendellösung" after the mechanical analogy with coupled pendulums. Malgrange and Authier (1965) performed an elegant experiment to demonstrate the complementarity of the intensities in the plane wave case. Using the double crystal spectrometer sketched in Fig. 1.7 they obtained a pseudo-plane wave of limited extent selected from the centre of the Borrmann fan and excited two tie points in the second crystal. This second crystal was, however, wedge-shaped and in the thin region, the wave-fields were not able to separate, and Pendellösung interference occurred. Fringes were thus seen in the diffracted and transmitted beams with complementary contrast. In the thicker regions, the wavefields diverged and interference could not take place. As predicted, no fringe contrast was then observed.

In the symmetric Bragg geometry, the z coordinate of Fig. 1.6b is the common coordinate. The angular deviation from the exact Bragg condition is then

$$\Delta\theta = -z \, \text{cosec} \, \theta_B / k. \qquad (1.56)$$

Substitution of (1.36) into (1.30) gives

$$0 = \Lambda_o^2 / 4 + \Lambda_o \eta \alpha_g \sec \theta_B + \alpha_g^2 \sec^2 \theta_B. \qquad (1.57)$$

This yields

$$\alpha_g \sec \theta_B = \tfrac{1}{2}\Lambda_o (- \eta \pm (\eta^2 - 1)^{\frac{1}{2}}) \qquad (1.58)$$

or

$$R = - \eta \mp (\eta^2 - 1)^{\frac{1}{2}}, \qquad (1.59)$$

where the + sign refers to branch 1 and the - sign to branch 2.

In the thin crystal case we can apply boundary conditions as in (1.44) and (1.48) and, as in the Laue case, interference effects occur. Bragg case

Pendellösung was observed experimentally by Batterman and Hildebrandt (1968)
and Nakayama, Hashizume, and Kohra (1970).

When the crystal is very thick a different situation arises. In Fig.
1.6b we note that the direction of energy flow of the tie point P_2 is outwards whereas that of P_1 is inwards. Now in a thick crystal, the amplitude
of the wavefield associated with P_2 must be zero if the crystal is thick and
even only moderately absorbing because such a wavefield may only reach the
crystal surface if it is created inside the crystal. In our case it could
only come from waves reflected from the back surface of the crystal, and if
the crystal is thick these are absorbed out. Thus the boundary conditions
become trivial,

$$D_o^e = D_o \text{ and } D_g^e = D_g \qquad (1.60)$$

and hence the intensity in the diffracted beam becomes

$$I_g = \{|\eta| - (\eta^2 - 1)^{\frac{1}{2}}\}^2 \qquad (1.61)$$

for a non-absorbing crystal. The intensity is thus independent of the
crystal thickness.

1.7.   RANGE OF BRAGG REFLECTION

As a crystal is rotated through the Bragg condition, the intensity, which
we have seen to be a function of η, the deviation parameter, varies. This
intensity versus angle curve is known as the rocking curve and, in general,
will be a convolution of the perfect crystal reflecting curve given by
(1.52) and (1.61) and the divergence of the X-ray beam. Let us suppose that,
by means of suitable monochromators, we have a beam of very low angular
divergence and highly monochromatic, i.e. a plane wave. The plane wave rocking curve in the symmetric Laue case with zero absorption is given by (1.52)
and is in general a rather complicated curve. However, in a thick crystal the
numerator, which varies rather rapidly with crystal thickness, may be ignored
and the intensity given by

$$I_g^\eta = (1 + \eta^2)^{-1}. \qquad (1.62)$$

This is plotted in Fig. 1.8a on the η scale, which is, of course proportional
to the angular deviation $\Delta\theta$. When absorption is included the curve becomes
asymmetric.

At intermediate thickness the numerator cannot be ignored, and
oscillations are seen in the wings of the curve. These Pendellösung effects
in plane wave rocking curves have been observed by Lefeld-Sosnowska and
Malgrange (1969) using the double crystal camera illustrated in Fig. 1.7.

The intensity of the curve in Fig. 1.8a drops to half its maximum value
at $\eta = \pm 1$. The full width of the curve at half height is thus 2 on the η
scale. Reference to (1.34) shows that this corresponds to an angle

$$\Delta\theta_{\frac{1}{2}} = 2/g\xi_g. \qquad (1.63)$$

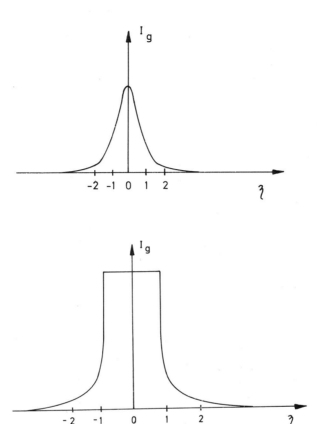

Fig. 1.8. (a) Perfect crystal reflecting curve for a moderately
          thick crystal in the Laue case. (b) Perfect crystal
          reflecting range for a thick crystal in the Bragg case.
          Over the range ± 1 on the η scale, the reflection is
          total.

Upon substituting approximate numbers we find that plane wave rocking curves
are of the order of a few seconds in width, i.e. 10^{-5} radians. As both g
and ξ_g rise with increasing order of reflection, high order, weak reflections
have extremely narrow reflecting ranges.
     The intensity in the Bragg case, given by (1.61), is plotted in Fig. 1.8b.
We note that for $|\eta| < 1$ the term under the square root is negative. The
reflecting power is unity and total reflection occurs. The curve for zero
absorption has the form shown in Fig. 1.8b, and again, absorption leads to
the curve becoming asymmetric. We can understand the significance of the
range of total reflection by reference to Fig. 1.6b. We see that $|\eta| = 1$
corresponds to the position where the normal no longer cuts the dispersion
surface. Thus for $|\eta| < 1$ no wavefields can be excited in the crystal, and

hence total reflection occurs. The range of total reflection in the symmetric Bragg case is given by (1.63).

1.8. GENERALIZED DIFFRACTION THEORY

The approach adopted in section 1.1 is successful in providing a clear physical insight into the diffraction processes occurring in a perfect crystal. However, it is not generally applicable to distorted crystals, and considerable effort was devoted in the late 1950s and 1960s to developing a theory capable of predicting diffraction contrast in distorted material. One of the most successful was the theory of Takagi (1962, 1969).

The wave inside the crystal is assumed to be composed of a wave of the form

$$\underline{D} = \exp(i\omega t) \sum_g \underline{D}_g(\underline{r}) \exp(-2\pi i \underline{K}_g \cdot \underline{r}) \qquad (1.64)$$

where

$$\underline{K}_g = \underline{K}_o - \underline{g} \quad \text{and} \quad |\underline{K}_o| = k(1 + \chi_o/2).$$

The important point is that we now have only one variable for each component wave $\underline{K}_g$. It cannot be stressed too strongly that this wavevector does not correspond to that of the wavefield as described in the previous sections. The wave amplitude $\underline{D}_g(\underline{r})$ is a function of position and therefore varies as the wave propagates. We will see that differential equations can be constructed to describe the changes and that these equations are very similar to those used in the Darwin treatment (Whelan, 1970). Substitution of (1.64) into Maxwell's equations and neglecting second order differentials leads to a set of equations

$$\frac{(\underline{K}_g \cdot \text{grad})\underline{D}_g}{k^2 - K_g^2} + i\pi \underline{D}_g + \frac{i\pi K_g^2}{k^2 - K_g^2} \sum_h \chi_{g-h} \underline{D}_h = 0 \qquad (1.65)$$

When we consider only one reciprocal lattice point we arrive at

$$(\hat{s}_o \cdot \text{grad})D_o = \partial D_o/\partial s_o = -i\pi k C \chi_{\bar{g}} D_g$$

$$(\hat{s}_g \cdot \text{grad})D_g = \partial D_g/\partial s_g = -i\pi k (C \chi_g D_o - 2\beta_g D_g) \qquad (1.66)$$

with $\hat{s}_o$ and $\hat{s}_g$ unit vectors in the incident and diffracted beam directions and

where $\beta_g = (K_g^2 - K_o^2)/2k = (K_g - k(1 + \chi_o/2))/k.$

Note again that $\underline{K}_g$ is not the same vector as in the classical dynamical theory. When $\beta_g = 0$, i.e. we have either a perfect crystal or a crystal containing a stacking fault, the equations may be solved analytically. However, when $\beta_g \neq 0$, numerical integration must be performed using a high-speed computer. Takagi's equations, as (1.66) are called, are an extremely powerful tool for calculating diffraction contrast in a highly distorted crystal, and we will see examples of their application in Chapter 3. The loss is that the concept of the dispersion surface is no longer retained, and it is extremely difficult to understand exactly what is going on inside the crystal. The development of a theory which retains a physical insight into the processes occurring inside the crystal is at present attracting much attention. Probably the most fundamental approach is that of Kuriyama. He has developed the quantum mechanical scattering matrix and derived the equation of diffraction in terms of it (Kuriyama, 1970). It can then be shown that the ray and wave theories which we discuss in Chapter 3 emerge as approximations of the general quantum mechanical theory (Kuriyama, 1972, 1973).

1.9.   EXTENSION TO ASYMMETRIC REFLECTION

The extension to cover asymmetric reflections is straightforward. The same procedures and criteria apply. All formulae, however, now contain terms in γ_o and γ_g, the cosines of the angles between the Bragg planes and the incident and diffracted beams respectively. The derivations may be found in the review articles so only results will be quoted.

For the asymmetric case, the equations for the intensity remain unchanged except that the parameters are defined differently. We find the deviation parameter now

$$\eta = k(\Delta\theta - \Delta\theta_o)\sin 2\theta_B/\Lambda_o|\gamma_g|$$

$$= \{\Delta\theta\sin 2\theta_B + \tfrac{1}{2}\chi_o(1 - \gamma_g/\gamma_o)\}/|C|(\chi_g\chi_{\bar{g}})^{\frac{1}{2}}(|\gamma_g|/\gamma_o)^{\frac{1}{2}}. \qquad (1.34A)$$

Here the minimum distance across the dispersion surface becomes

$$\Lambda_o = kC(\chi_g\chi_{\bar{g}})^{\frac{1}{2}}/(|\gamma_g|\gamma_o)^{\frac{1}{2}}, \qquad (1.41A)$$

i.e. the extinction distance becomes

$$\xi_g = (|\gamma_g|\gamma_o)^{\frac{1}{2}}/kC(\chi_g\chi_{\bar{g}})^{\frac{1}{2}} = (|\gamma_g|\gamma_o)^{\frac{1}{2}}\pi V_c/r_e\lambda C(F_g F_{\bar{g}})^{\frac{1}{2}}. \qquad (1.54A)$$

The full width at half height of the reflecting range is now given by

$$\Delta\theta_{\frac{1}{2}} = 2|\gamma_g|/\cos\theta_B g\xi_g = 2|C||\chi_g|(|\gamma_g|/\gamma_o)^{\frac{1}{2}}/\sin 2\theta_B \qquad (1.63A)$$

We note that the half width is dependent on the ratio $(|\gamma_g|/\gamma_o)$, and thus at grazing incidence the reflecting range can become extremely narrow. This is particularly important in the Bragg case, and has been exploited in the work on asymmetrically cut monochromators by Kohra's group described briefly in the next chapter.

1.10. <u>ANALYSIS</u>

We have now developed the theory of diffraction in perfect crystals and attempted to see how electromagnetic waves behave in such crystals. While important to the X-ray topographer, such theory must be regarded as a fundamental background as it is with imperfect crystals that he is concerned. The rigorous approach to the imperfect crystal has been indicated in section 1.8. However, we will not continue this rigorous treatment, and in Chapter 3 we will use the results obtained in the perfect crystal to obtain a qualitative understanding of the contrast obtained in topographs of imperfect crystals. The concepts of the dispersion surface and Bloch wavefields will figure strongly in the discussion. Before considering the contrast in the topographs though, it is desirable to examine the more common experimental arrangements for taking the topographs themselves. One of the most importance features to emerge will be that in almost all topographic apparatus the incident beam cannot be represented by a plane wave.

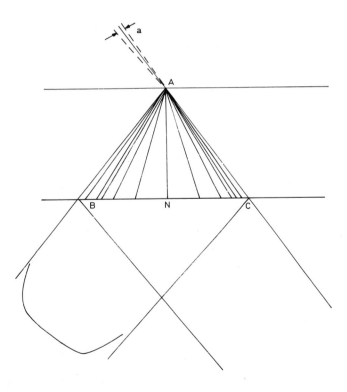

Fig. 1.9. Density of wavefields within the Borrmann fan for an incident
          spherical wave where all tie points along the dispersion
          surface are excited.

In practice, it is found that the significant angular divergence of the wave causes the whole of the dispersion surface to be excited simultaneously. This causes energy to flow in a range of directions in the crystal. The whole of the dispersion surface is then excited and the whole region within the triangle BAC of Fig. 1.9 is filled with wavefields propagating in different directions. The significance of this filling of the Borrmann triangle will be investigated at the beginning of Chapter 3.

References

Authier, A. (1960) *C.R. Acad. Sci. Paris* <u>251</u>, 2003.
Authier, A. (1961) *Bull. Soc. Fr. Min. Crist.* <u>84</u>, 51.
Authier, A. (1970) In *Advances in Structure Research by Diffraction Methods*
          <u>3</u> (ed. Brill and Mason), p.1.
Authier, A., Balibar, F., and Epelboin, Y. (1970) *Phys. Stat. Sol.* <u>41</u>, 225.
Batterman, B. W. and Cole, H. (1964) *Rev. Mod. Phys.* 36, 681.
Batterman, B. W. and Hildebrandt, G. (1968) *Acta. Cryst.* A24, 150.
Borrmann, G., Hartwig, W., and Irmler, H. (1958) *Z. Naturforschung* <u>13a</u>, 423.
Dederichs, P. H. (1970) *Phys. Rev.* B <u>1</u>, 1306.
Fathers, D. J. and Tanner, B. K. (1973) *Phil. Mag.* <u>28</u>, 749.
Hart, M. (1971) *Rept. Prog. Phys.* <u>34</u>, 435.
Hart, M. and Milne, A. D. (1971) *Acta Cryst.* A27, 430.
James, R. W. (1948) *The Optical Principles of the Diffraction of X-rays,*
          Bell, London.
Kato, N. (1963) In *Crystallography and Crystal Perfection* (ed. Ramachandran),
          p. 153, Academic Press.
Kuriyama, M. (1970) *Acta Cryst.* A26, 56.
Kuriyama, M. (1972) *Acta Cryst.* A28, 588.
Kuriyama, M. (1973) *Z. Naturforschung* 28a, 622.
Laue, M. von (1952) *Acta Cryst.* <u>5</u>, 619.
Laue, M. von (1960) *Rontgenstrahlinterferenzen*, Akad. Verlagsgesellschaft,
          Frankfurt.
Lefeld-Sosnowska, M. and Malgrange, C. (1969) *Phys. Stat. Sol.* <u>34</u>, 635.
Malgrange, C. and Authier, A. (1965) *C.R. Acad. Sci. Paris* <u>261</u>, 3774.
Meieran, E. S. and Blech, I. A. (1968) *Phys. Stat. Sol.* <u>29</u>, 653.
Nakayama, K., Hashizume, H., and Kohra, K. (1970) *J. Phys. Soc. Japan* <u>30</u>,
          893.
Nøst, B., Larson, B. C., and Young, F. W. Jr. (1972) *Phys. Stat. Sol. (a)*
          <u>11</u>, 263.
Patel, J. R. (1973) *J. Appl. Phys.* <u>44</u>, 3903.
Takagi, S. (1962) *Acta Cryst.* <u>15</u>, 1311.
Takagi, S. (1969) *J. Phys. Soc. Japan* 26, 1239.
Whelan, M. J. (1970) In *Modern Diffraction and Imaging Techniques in*
          Material Science (ed. Amelinckx et al.), p.35,
          North-Holland.
Zachariasen, W. H. (1945) *Theory of X-ray Diffraction in Crystals,*
          Wiley, New York.

CHAPTER 2

EXPERIMENTAL TECHNIQUES

2.1.   PRINCIPLES

Historically it is hard to identify exactly when X-ray topography was
invented as present day techniques are a result of a slow evolution culmina-
ting in the transmission technique devised by Lang (1958) and the double
crystal technique devised independently by Bond and Andrus (1952) and Bonse
and Kappler (1958). While the double crystal method has recently become
extremely important in the study of device materials, its limited application prior to the mid-sixties may have arisen from the relative imperfection
of the available crystals. As a result it was the work of Lang which
stimulated the growth of topography as his technique, though having high
spatial resolution was less strain sensitive and thus was applicable to the
study of more crystals.

It is not my aim to provide a detailed review of the historical development of topography or a description of the many modified techniques which
have been reported over the years. Most are included in the appendix and
are discussed in the review article by Lang (1970). Briefer reviews of
techniques have been given by Bonse, Hart, and Newkirk (1967) and Isherwood
and Wallace (1974). Here, three techniques in current usage will be
described in order of sensitivity. They may thus be considered in the order
relevant to the crystal grower at successive stages in a programme to
produce highly perfect single crystals. The methods are

    (1)   Berg-Barrett Method
    (2)   Lang's Method
    (3)   Double Crystal Methods

The aim of all X-ray topographic methods is to provide a picture of the
distribution of the defects in a crystal, and X-ray images may be thought of
as arising in two ways. These are ORIENTATION contrast and EXTINCTION
contrast. Suppose the crystal is set in an X-ray beam of divergence $\Delta\phi$ so
that Bragg reflection from a particular set of lattice planes is achieved
for one or more characteristic lines. Then if the diffracted beam is
examined as a function of position in the crystal, we might expect a perfect
parallel-sided crystal to present a uniform field of view, and, indeed, this
is what is found in a projection topograph. Should the crystal contain a
region, e.g. a twin, misoriented with respect to the otherwise perfect
crystal, contrast between this region and the crystal matrix will be observed
if the misorientation exceeds $\Delta\phi$. In this case Bragg reflection of the
characteristic lines cannot take place in the misoriented region, leaving an
undarkened patch on the film. (Fig. 2.1a). This is orientation contrast, and
it can be interpreted quite simply geometrically without recourse to

24

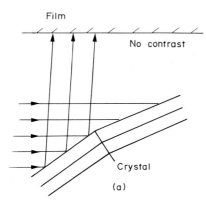
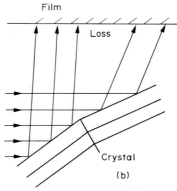

Fig. 2.1. (a) Orientation contrast from a monochromatic collimated
X-ray beam. Diffraction occurs only from one region.
(b) Orientation contrast from a polychromatic X-ray beam.
The differing directions of the beams from the two regions
leads to contrast at the boundary if the plate - specimen
distance is sufficiently large.

dynamical theory.
    Orientation contrast may also arise from the different directions of the
beams diffracted by misoriented regions. If the film is placed a considerable distance from the specimen, spatial overlap or separation occurs leading to enhanced or reduced intensity corresponding to the boundaries of the
misoriented region (Fig. 2.1b). In this way, orientation contrast may be
obtained using continuous radiation and this forms the basis for the
experimentally simple but little-used Schulz technique (Schulz, 1954). The
unpopularity is probably due to difficulties encountered in interpreting
Schulz topographs which are easy to understand only when the crystal consists of discrete mosaic blocks (Aristov and Shulakov, 1975). Complex
images result when the lattice distortion is continuous. For high angular
sensitivity, the specimen to film distance must be long and, therefore,
unless a microfocus tube is used, the spatial resolution is poor. Although
a white radiation technique, it is sensitive to extinction contrast as well,
and in the transmission setting is gaining in popularity due to the advent
of synchrotron radiation (Tuomi, Naukkarinen, and Rabe, 1974).
    Extinction contrast arises from the distortion of the lattice around a
defect giving rise to different diffraction conditions from those in the
surrounding matrix. There is thus a different scattered intensity from the
vicinity of the defect. The nature of the image can only be deduced using
dynamical diffraction theory, and this will be discussed in detail in
Chapter 3.

2.2. THE BERG-BARRETT METHOD

    This is one of the older methods originating from improvements by
Barrett (1945) on earlier work by Berg (1931). It is most commonly used in
reflection where it has the virtue of being applicable to quite high dislocation density materials. In the production of new crystals it might be
considered the first assessment technique to apply.

The essential features of the Berg–Barrett method are shown schematically
in Fig. 2.2. The crystal is set to obtain diffraction from a particular set
of lattice planes of a characteristic line coming from an extended source.

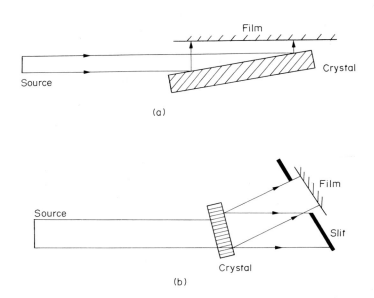

Fig. 2.2. (a) Reflection Berg–Barrett technique of topography.
          (b) Transmission Berg–Barrett (Barth–Hosemann) technique.

A photographic plate placed some distance from the crystal is used to record
the diffracted beam. Due to the use of an extended source and small specimen to plate distance, it is relatively insensitive to orientation contrast,
though large misorientations of the order of a degree will result in no
diffracted intensity reaching the plate. The method is appealing as no
expensive or high-precision components are required. Most important is that,
because of the wide source, the position of the crystal is not critical, and
Bragg reflection is obtained within an angular range of typically 1°
(10^{-2} rad). This wide "rocking curve" makes setting up very easy. The
major problem, the loss of resolution due to the polychromaticity of the
X-ray beam, is overcome in different ways depending whether in reflection or
transmission.

2.2.1. Reflection

The big problem in X-ray topography is that the $K\alpha$ line is in fact a
closely spaced doublet where the intensity ratio of the $K\alpha_1$ to $K\alpha_2$ lines is
2 : 1. This results in a doubling of the image of any defect due to the
different diffraction angles of the two lines. Barrett overcame this
problem in reflection by using a smaller source, less than about 1 mm, with
a crystal cut so that the required reflecting planes were not parallel to the

surface. Careful selection of specimen geometry enables asymmetric reflections to be used where a wide beam is obtained from a relatively narrow one.
It is also possible to arrange for the diffracted beam to be directed almost
normal to the specimen surface. Under such conditions it is possible to
place the film very close to the crystal surface as shown in Fig. 2.2a.
The divergence of the $K\alpha_1$ and $K\alpha_2$ lines is thus avoided, leading to narrow,
single images of the dislocations. Of course, when the film is so close
to the specimen surface it must be nearly parallel to it and cannot always
be perpendicular to the diffracted beam as would be ideal. There may thus
be a loss of resolution due to the X-rays passing through the film at an
angle, and this can be minimized by using very thin photographic emulsions
(10 μm). For an appreciable fraction of the X-rays to be absorbed it is
necessary to use soft radiation, but this is often convenient as the large
Bragg angles then encountered make it easy to find low-order asymmetric
reflections which give optimum resolution conditions.
    An important feature of the use of soft radiation is that the extinction
distance is very small and the diffracted X-rays penetrate only a very small
distance into the crystal. We are then examining only a very thin slice of
the crystal close to the surface; the rest of the crystal is irrelevant to
our experiment. As the ability to resolve individual dislocations in a
topograph is limited by the overlapping of defect images at different depths
in the crystal, it is seen that reflection rather than transmission techniques are applicable to high dislocation density crystals. The Berg-
Barrett method in reflection is useful up to dislocation densities of about
10^6 cm^{-2}.
    It is usual to use a counter placed in the 2θ position to detect a strong
Bragg reflection when the crystal is slowly rotated in the horizontal plane
but it is possible - though not desirable - to find the reflection by looking for the diffracted beam on a phosphor screen. Under blackout conditions
this can be observed without image intensification, but there is a serious
radiation hazard attached to such practice. An excellent review of the
reflection Berg-Barrett method, including the design of a simple camera, has
been given by Austermann and Newkirk (1967). Details of more sophisticated
camera design can be found in the paper by Liang and Pope (1973), and many
useful operating hints have been recorded by Turner, Vreeland, and Pope
(1968). A method of analysis of grain boundary contrast based on the
stereographic projection has been developed by Wu and Armstrong (1975).

2.2.2. Transmission (Barth-Hosemann Geometry)

    In transmission (Fig. 2.2b) it is not normally possible to place the film
close to the specimen surface as this results in overlapping of the direct
and diffracted beams. The exception to this is the case of a very thick
crystal where only X-rays very close to the Bragg reflection conditions are
transmitted. In this anomalous transmission situation, defects cause loss
of intensity in both direct and diffracted beams. The film may then be
placed in contact with the exit face of the crystal, and studies of dislocations and domains using this technique have been performed by Roessler
and colleagues on Fe-Si alloys (Roessler 1967; Kuriyama and McManus, 1968).
    When the crystal is thin and the direct beam contains a very high
intensity of undiffracted radiation the film and specimen must be separated
to allow the stopping of the direct beam by a slit. Several techniques have
been devised to circumvent the dispersion problem which marred the original
work of Barth and Hosemann (1958).

(1) Use of the Ag $K\alpha_2$ line when the $K\alpha_1$ component has
been removed by a Ru foil filter (Hosoya, 1968).

(2) Use of a Soller slit to reduce the beam divergence
and oscillating it to remove the stripes from the
X-ray film (Oki and Futagami, 1969).

(3) Use of the $K\beta$ line (Dionne, 1967). The $K\beta_1 : K\beta_2$
intensity ratio is about $100 : 1$, and as the $K\alpha$
and $K\beta$ are separated by about 1° it is easy to
obtain single images. The main disadvantage of the
method is that the $K\beta$ line is considerably weaker
than the $K\alpha$ line.

The transmission Berg-Barrett method is particularly susceptible to a high
background of scattered radiation. Extreme caution should be exercised both
with respect to the risk of irradiation and the signal-to-noise level on the
micrograph. While the former trouble can be eliminated by the use of well-
constructed lead shielding, the latter represents the greatest fault of the
transmission Berg-Barrett method. This disadvantage is offset by the low
sensitivity to orientation contrast, valuable when examining bent crystals,
and the simplicity of the apparatus. The scattered radiation leads to a
loss in contrast which is particularly bad on the dynamical image but leads
to no appreciable loss of resolution in the direct image which is comparable
to the resolution of Lang's method.

2.3.   LANG'S TECHNIQUE

This is the most popular and widely used technique of transmission X-ray
topography, and as such it will be considered in some detail. It is
sensitive to both extinction and orientation contrast, the orientation
sensitivity being about 5×10^{-4} rad. Double images arising from simultaneous reflection of the $K\alpha$ doublet are avoided by collimating the beam to
such an extent as to allow diffraction from one $K\alpha$ line only at once. The
$K\alpha_1$ line is used due to its greater intensity. We note that the typical
beam divergence in the Lang method is 5×10^{-4} rad, well over an order of
magnitude greater than the perfect crystal reflecting range of about 10^{-5}
rad. It is straightforward to show that the contribution to the image of
rays between 5×10^{-4} and 10^{-2} rad away from the exact Bragg condition is
negligible, and thus images in the Lang and transmission Berg-Barrett
method are almost identical except for the background scatter. "Projection
topographs" of the whole crystal are formed (Lang, 1959) by traversing the
crystal and film together across the beam (Fig. 2.3). Integration in time
yields a topograph equivalent to the one taken by the Berg Barrett method
where the integration is in space.
However, the projection topograph throws away information available in the
stationary or "section" topograph formed when the beam is narrow compared
with the base of the Borrmann fan (Lang, 1958), and thus the section topograph should be regarded as the more fundamental. Scanning of a section
topograph of course results in the projection topograph again. Referring to
Fig. 1.9, when the width of the incident beam is small compared with the
width of the Borrmann fan, i.e.

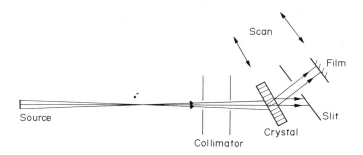

Fig. 2.3. Lang's transmission technique. Section topographs
          correspond to the stationary situation, projection
          topographs are taken by scanning crystal and film across
          the beam.

$$a \ll 2t \sin \theta_B, \tag{2.1}$$

the image records information on the direction of energy flow within the
crystal. In very thin crystals and with small Bragg angles one can often not
obtain this information, however, as it is extremely difficult to meet
condition (2.1) experimentally. It is often convenient to stop at a point on
a projection topograph and examine the small region chosen with the aid of a
series of section topographs. For this one needs only to modify the collimating slits of the Lang camera. A slit of 10 μm is generally employed compared with slits up to several hundred microns for projection topographs.

      Section topographs are extremely valuable in determining the depth of a
defect in a crystal, and this facility has been exploited by Gerward and
Lindegaard Anderson (1974) in step-scanned section topographs. An alternate
method for determining the position of defects is to take stereo pairs of
projection topographs. This is less sensitive and only qualitative, but is
quick to perform and enables a large area to be scanned at a time. Stereo
topographs are often taken as a pair in the hkl and h̄k̄l̄ reflections, and for
low absorbing crystals this works well. However, under moderate absorption
conditions, Friedel's law breaks down, and the contrast of many defects
reverses between hkl and h̄k̄l̄ reflections. It is then preferable to use a
single reflection but to take two topographs rotated ±8° about the diffraction vector(Haruta, 1965). In both cases, the stereo effect increases with
crystal thickness and good stereo pairs cannot be obtained from thin crystals.

      Lang cameras are now commercially available but home construction is not
too difficult, the main problem is the construction of an accurately traversing table on the rotational axis. The traverse unit is critical to the
success of all subsequent experiments, and common features on projection
topographs are traverse lines or striations. These usually result from
periodic fluctuations in the traversing speed with a frequency corresponding
to the pitch of the driving screw. Certain areas are then exposed longer
than others leading to VERTICAL stripes across the image. Almost all
cameras exhibit traverse lines to some extent, on the better ones the lines
are not visible in the final reproduction.

Fig. 2.4. Photograph of the Lang camera constructed in Durham
        University Physics Department.

Fig. 2.4 shows a photograph of the Lang camera built at the Physics
Department of Durham University to what is essentially Lang's original design.
Many features are similar to the camera now marketed by Marconi-Elliott. The
specimen is mounted on a small goniometer with two axis rotation and preferably two degrees of lateral movement. This latter adjustment is particularly
valuable when taking topographs of small crystals, as it is then essential to
centre the crystal above the main rotational (θ) axis when searching for the
Bragg reflection. When this is not done, fiddly adjustment of the table
position using a fluorescent screen is necessary to ensure that the crystal
remains in the beam upon rotation.
    A suitable collimator buts into the surround of the output window in the
X-ray tube housing. Considerable care should be taken to ensure that no
radiation escapes as the flux is very high from modern fine focus units and
radiation leakage is prone to occur in very narrow beams of high intensity.
It is therefore vital that the tube housing and collimator be compatible, and
experimenters are strongly advised to make their own adaptor to fit their
particular X-ray set. Wrapping lead sheet around an incompatible union will
not do. Periodic checks with a radiation monitor of the union are strongly
advised.
    The collimating slits c should also be adequately shielded and care taken
to ensure no radiation leaks from the junction of the collimator tube and the
slits. Here the X-ray intensity is much less than at the union with the tube
housing and the hazard correspondingly less. Slits are preferably double to

reduce scatter and are usually manufactured from tantalum which gives high
absorption of X-rays while being easily machinable. The diffracted beam
slits s are similarly machined out of tantalum and the slit width and position can be varied by means of micrometers. Precise position is achievable,
necessary for the technique of <u>limited projection topography</u> (Lang 1963).

The technique was originally devised for use in studies on natural
diamonds where it was impossible to remove the strained surface layer
resulting from mechanical damage. Usually, it is possible to remove such
damage by chemical or electrochemical etching, but in the particular case of
gemstones the owners often require them for other purposes. When damage is
not removed, the contrast from it obscures any weaker images from the interior defects. Limited projection topography enables this contrast to be
eliminated in part. The diffracted beam slits (Fig. 2.5) are arranged to
cut into the beam and allow only part of the diffracted beam to reach the

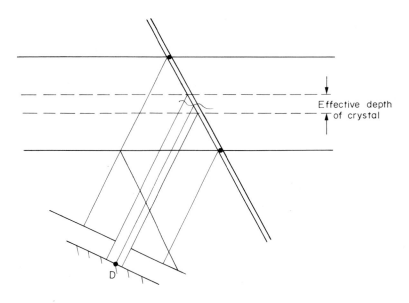

Fig. 2.5. The limited projection technique. The narrow diffracted
          beam slits select only information from a particular depth
          of crystal.

film. Under conditions of low absorption, such as exist in diamond, most of
the intensity in the image results from the scattering out of the direct
beam X-rays not exactly satisfying the Bragg condition. (See Chapter 3).
This "direct" image is formed at a point across the diffracted beam
corresponding to the depth of the defect in the crystal. As illustrated
in Fig. 2.5, most of the surface images can be removed or, alternatively,
only images formed from defects within a small distance of the surface
imaged. The latter application was useful in establishing the one to one

correspondence of etch pits and dislocation outcrops. While ingenious, the
technique has limited application as it can only be used in conditions of
low absorption and where the width of the base of the Borrmann fan is
substantial. It is not a particularly sensitive method for determining the
depth of defects in the crystal.

   The table t is rotated on a central spindle mounted on   standard steel
ball journal bearings. It is traversed on PTFE balls by means of a micrometer driven either by Selsyn torque transmitters or a stepper motor directly
on the table support. Use of a stepper motor for driving has the advantage
that a simple binary counter can be used to control the length of traverse
and is easier to adjust than mechanical limit switches. Control of the
specimen position in the horizontal plane is achieved by a 50 mm micrometer
pushing on a lever l. Detection of the diffracted beam is by a NaI (Tℓ)
scintillation crystal backed by a photomultiplier. This assembly p is
mounted on an arm at right angles to the diffracted beam slits and the two
arms rotate together, although facility is provided to adjust the angle
between them and the counter window size if required. It is possible to use
a Geiger counter or even a small radiation monitor to detect the beam, but
response is slow and Geiger counters saturate at high count rates. The
improved results with a scintillator are worth the expense.

2.4. EXPERIMENTAL PROCEDURES FOR TAKING LANG TOPOGRAPHS

   In experimental science there is no substitute for experience, but the
way of the novice can often be smoothed if certain procedures are followed
and adapted. The following sections are written for the benefit of beginn-
ers in topography, and old hands are welcome to skip them. Some of the
material here can already be found in the literature (e.g. Lang, 1970;
Austermann and Newkirk, 1967), but some things are new or exist only in the
folk-lore. I hope that anyone entering the field will find the next
sections useful.

2.4.1. Setting up the Crystal

   An important feature, often overlooked, is the technique of mounting the
crystal. When the crystal is small, it is useful to mount it inside a
suitable metal ring which has a number of holes drilled in it for fixing to
the goniometer. This enables a large number of reflections to be taken
simply by rotating the mount and not having to interfere with the crystal.
X-ray topographic techniques are necessarily sensitive to strain, and
inconsiderate mounting will lead to difficulties. In plastic crystals many
dislocations can be introduced and in brittle materials a radius of curvature as great as 2 m will prevent Bragg reflection from occurring simultaneously across a 1 cm wide sample. For many purposes a soft wax is a suitable mounting material provided that only the minimum necessary to hold the
crystal is used. (We use Cenco Soft-Seal wax and melt it by gently heating
with a small soldering iron which we are careful not to overheat.) Fixing
at one point only is recommended as two blobs of wax will often strain the
crystal. Quick-drying glues which contract on setting should be avoided.
Where a permanent job is required an epoxy resin or strain gauge cement can
be successfully used as they contract very little on setting and a general
rule is to apply the adhesive to the sample as far away from the region
under examination as possible. At low temperatures a small quantity of

varnish makes a useful adhesive.

2.4.2. Setting the Diffraction Vector in the Horizontal Plane

In crystals conveniently displaying facets or edges oriented in definite
crystallographic directions the problem is simple. The crystal can be
aligned by eye using a draughtman's variable angle set square. For Lang
topography this will be quite adequate, though fine adjustment can be made
once the Bragg reflection has been found. Where the crystal has not
crystallographically oriented edges, the whole goniometer should be trans-
ferred to a back-reflection Laue camera and the crystallographic orientation
determined. Suitable rotation will then bring the required diffracting
planes into a vertical orientation.

2.4.3. Finding the Bragg Reflection

When setting up the experiment, the use of low X-ray power is advised to
reduce the irradiation risk. High voltage-low current conditions give a
good characteristic line intensity, and care should be taken to ensure that
the critical voltage is exceeded. (You may have a long hunt for the Bragg
reflection with AgKα at 20 kV). Beam monitoring is best done in a semi-
darkened room using a small fluorescent screen. This is conveniently made
by sticking a fluorescent powder onto a thin piece of tufnol or perspex and
should have a handle long enough to remove fingers from the beam area. The
thinner the backing, the easier it is to determine the position of the
specimen relative to the beam. A good practice is to run the crystal into
the beam holding the fluorescent screen behind the crystal and then to use
the photomultiplier.

As the crystal may not be centred over the camera axis, it is advisable to
reduce the distance between specimen and scintillator for initial setting.
With a non-centred specimen, the effective 2θ angle of the counter will not
be given by $2\theta_B$ read on the counter scale. A wide angular aperture avoids
trouble but increases the background. Once the reflection is found, the
angular acceptance should be decreased and the scintillator moved further
from the specimen. This eliminates the possibility of having found a stray
reflection from non-vertical diffraction planes. The crystal should be
turned slowly, checking that the crystal remains in the beam. Some days it
is easy, some days it is not.

Once the peak is found, the coarse adjustment should be locked and the
fine adjustment used to locate the Kα_1 peak. Two peaks should be visible,
the Kα_2 being half of the intensity of the Kα_1. If this intensity ratio is
not 2 : 1 it implies that the Bragg planes are not vertical and the orientation of the crystal in the plane normal to the incident beam should be
adjusted. A goniometer key mounted on a 20 cm rod enables the crystal to be
oriented under X-ray illumination without exposing the fingers. Some
commercial cameras have a rotational facility built into the goniometer.

When anomalous transmission studies are made of materials giving
appreciable fluorescence it is sometimes impossible to detect a Bragg
reflection from the thick crystal against the high background. In this situation the very intense surface reflection from the edge of the crystal can be
used to great advantage. The crystal should be adjusted until it just cuts
into the X-ray beam as observed on the monitoring screen. Once the edge

reflection is found, the crystal can be traversed into the thick region and
the low visibility Bragg peak there will be seen.   In the event of no peak
being observed one can conclude fairly certainly that anomalous transmission
is not occurring.

Once the reflection has been found, the diffracted beam slits may be
inserted.   With low fluorescence it may be satisfactory to use only a
straight edge to eliminate the direct beam, but in many cases the use of a
slit is required.   At this point, one may assess the quality of the crystal.
A sharp rocking curve (intensity versus horizontal setting) indicates that
the crystal is not warped and has a relatively low dislocation density.
A broad rocking curve in which it is impossible to distinguish the $K\alpha_1$ and
$K\alpha_2$ components is a good indication that your crystal is highly strained and
your topograph will not be very beautiful.   Similarly, a rapid change in
intensity on traversing, giving a strong signal over only a short range, is
often indicative of a bent crystal.

In the last circumstance, provided the curvature is not too great and
hard radiation is used, it is possible to obtain good topographs by continuously adjusting the angular setting of the crystal to keep it exactly on the
Bragg reflection.   This can be done by hand, but for exposures over an hour
it becomes tedious.   Several devices have been developed to perform the
operation automatically (Angillelo and Wood, 1971; van Mellaert and
Schwuttke, 1970).   At least one device is available commercially from
Precision Devices, Malvern, UK, having being developed by Hart's group at
Bristol.   The principle is extremely elegant.   The crystal sits on a piezoelectric transducer fed with an a.c. signal which causes the crystal to
oscillate with an amplitude up to 1 minute of arc.   Consequently the diffracted beam intensity varies.   The signal onto the piezoelectric and the output
from the photomultiplier are fed into a phase-sensitive detector, the output
of which is then either positive, zero, or negative depending on the position
of the crystal on the rocking curve (Fig. 2.6).   This signal is used to drive
a reversable d.c. motor on the crystal axis.   The method is much faster than
the original scanning oscillation technique of Schwuttke (1965), less
expensive than the device of van Mellaert and Schwuttke and more flexible
than that of Angillelo and Wood.

2.4.4. Recording the Topograph

As will become apparent, it is important to place the photographic plate
as close to the specimen as possible.   With rotating anode generators, care
should be taken not to allow the full power of the beam to fall on the plate
when stationary as this leads to an unsightly overexposed vertical line on
the topograph.   Slow traverse speeds are recommended for exposure, typically
10 cm/hour.   This strikes a balance between a sufficiently rapid rate to
provide multiple passes to average out intensity fluctuations and sufficiently slow rate to avoid disturbing the crystal and film cassette by violent
reversals of the traverse direction.

When using hard radiation it is useful to check the setting at least once
during the exposure through the back of the plate.   Small drifts will not
seriously mar the topographic resolution but will lead to under exposure.

The presence of a horizontal stripe on the recorded topograph sometimes
gives rise to concern and this is often due to the presence of a second
reciprocal lattice point lying on the Ewald Sphere.   It can often be removed
by a small rotation of the crystal about the diffraction vector as if to take

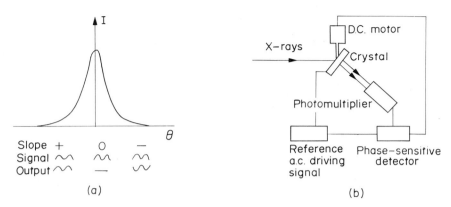

Fig. 2.6. (a) Rocking curve showing how the sign of the gradient
         may be determined by the relative phase of an oscillatory
         input and output. (b) Schematic diagram of a Bragg angle
         controller.

a stereo pair.

2.5.   TOPOGRAPHIC RESOLUTION

   As the criteria differ in the horizontal and vertical directions, they
will be considered separately.

2.5.1. Vertical Resolution

   In Lang topography, the diffracting planes are vertical so there is
essentially no diffraction in the vertical plane. The resolution, δ, on the
plate is given by Fig. 2.7.

$$\delta = LV/D, \tag{2.2}$$

where L is the specimen-plate distance, D the source - specimen distance and
V is the projected height of the source. As it is practically impossible to
reduce L below 1 cm, when one substitutes typical values, a resolution in the
order of a few microns is obtained. A small specimen-plate distance should
always be aimed for when setting up a topograph.

2.5.2. Horizontal Resolution

   The divergence of the diffracted beam in the horizontal is very small and
is not the limiting factor on resolution. Four factors merit consideration.

(a)   Vibration

      It is vital to ensure that the specimen and plate do not move relat-
ive to one another. When a continuously pumped generator is used, the pumps

Fig. 2.7. Vertical resolution criteria.

should be removed from the chassis and often it is desirable, particularly
with a rotating anode generator, to mount the camera on an independent heavy
table. A simple antivibration mount consists of a piece of thick foam
rubber beneath the camera.

    (b)   Fundamental width of the X-ray line

        The spectral width of the characteristic X-ray lines, although small,
can under some circumstances lead to a loss of resolution. Differentiation
of the Bragg relation shows the spread in Bragg angle $\Delta\theta_B$ to be related to
the wavelength spread $\Delta\lambda$ by

$$\Delta\theta_B = \tan\theta_B \, \Delta\lambda/\lambda. \tag{2.3}$$

This will be more important for large Bragg angles than for small ones.
Table 2.1 illustrates this.

TABLE 2.1. Effect of natural line width on the X-ray resolution for a specimen to plate distance of 1 cm

	CuKα$_1$ $\Delta\lambda/\lambda = 4 \times 10^{-4}$			AgKα$_1$ $\Delta\lambda/\lambda = 5 \times 10^{-4}$		
$d(\text{Å})$	θ_B^o	$\Delta\theta\alpha_1$ (sec)	Δx (μm)	θ_B^o	$\Delta\theta\alpha_1$ (sec)	Δx (μm)
3.13	14	20	1	5	9	0.4
1.92	24	36	1.7	8.5	15	0.7
1.1	44	81	3.9	14	27	1.3

Generally this contribution is small, as large Bragg angles are rarely used.

(c) <u>Simultaneous diffraction of both the Kα lines</u>

Closely related to the previous consideration is the problem of diffraction of more than one line. Substitution into (2.3) shows that the angular separation of the diffracted beams due to the $K\alpha_1$ and $K\alpha_2$ lines is between 10^{-4} and 10^{-3} rad. Each point in the image is thus displaced by $L\Delta\theta_{\alpha_1\alpha_2}$ and double images result. We note from Table 2.2 that the separation is significant even for the smallest obtainable specimen to plate distance.

TABLE 2.2. Separation of the $K\alpha_1$ and $K\alpha_2$ images for various reflections in silicon. Specimen to plate separation is 1 cm

	$\Delta\lambda_{\alpha_1\alpha_2}$ (Å)	111 reflection $\Delta\theta$ (rad)	Δx (μm)	220 reflection $\Delta\theta$ (rad)	Δx (μm)
CrKα	3.9×10^{-3}	7×10^{-4}	7	1.3×10^{-2}	13
CuKα	3.8×10^{-3}	6×10^{-4}	6	1.0×10^{-3}	10
MoKα	4.3×10^{-3}	7×10^{-3}	7	1.1×10^{-3}	11
AgKα	4.4×10^{-3}	7×10^{-4}	7	1.2×10^{-3}	12

The maximum permissible beam divergence is that which just allows the resolution of the $K\alpha_1$ and $K\alpha_2$ peaks in the rocking curve; if too great a separation is used we will see later that the exposure time is being unnecessarily increased.

(d) <u>Fundamental width of dislocation images</u>

Images in X-ray topographs are two to three orders of magnitude wider than those in transmission electron micrographs due to the corresponding difference in the angular width of the reflecting range around reciprocal lattice points. For X-rays this is typically 10^{-5} compared with 10^{-2} rad for 100 KeV electrons. Equation (1.63) shows that the rocking curve width is inversely proportional to the extinction distance, which measures the strength of the reflection. The extinction distance is inversely proportional to the scattering factor, and hence it is the relatively weak scattering of X-rays by atoms that leads to the narrow reflecting range.

We can now qualitatively understand the reason for the wide images in X-ray topographs. Images are formed when the lattice planes are misoriented so that perfect crystal reflection cannot take place. The exact criteria for image formation are not important here, but we note that the misorientation around a dislocation line varies roughly as r^{-1}. The narrower the reflection range the further from the dislocation core will the crystal appear "misoriented". Roughly, we may expect dislocation images in X-ray topographs to be two or three orders of magnitude greater than in electron micrographs, which is what is observed.

High order, weak reflections have very narrow reflecting ranges and hence high sensitivity to lattice misorientation. This is exploited in some of the recent double crystal work. Thus weak, high-order reflections will give wide dislocation images. For maximum spatial resolution, strong low-order reflections should be used. The half width of the reflecting range is, for symmetric reflections,

$$\Delta\theta_{\frac{1}{2}} = 2Cr_e\lambda(F_{\bar{g}}F_g)^{\frac{1}{2}}/\pi gV_c \cos\theta_B, \tag{2.4}$$

and thus is proportional to wavelength. Therefore the width of dislocation images increases with wavelength. For optimum resolution, long-wavelength low order strong reflections should be employed, and Lang's topographs of dislocations in iron silicon are good examples of what can be achieved under optimum conditions. Images as narrow as 1 μm have been recorded.

2.6. PHOTOGRAPHY

Unlike the electron microscopist, the X-ray topographer has no magnification in his X-ray image. The topograph must be recorded at a magnification of unity and subsequently enlarged optically. As dislocation images are upwards of a micron in width it is essential that a recording medium of comparable resolution is used. It is widely accepted that the Ilford L4 Nuclear Emulsions (undeveloped grain size 0.14 μm) have excellent characteristics for the recording of X-ray topographs. G5 emulsions, although faster, have larger grain size, and use is not recommended as information can easily be lost from the topograph. Dental film is fast and can be used to obtain a large area image, but as the resolution is poor this should normally be regarded as a trial exposure. In the optical case a very thin emulsion would be chosen to obtain high resolution, but due to the relatively poor absorption of X-rays there is a limit to the minimum thickness tolerable. For harder radiations, e.g. MoKα and AgKα, 50 μm thick emulsion is used while for softer radiation 25 μm thickness is satisfactory. These constitute a compromise between efficiency and resolution, a dilemma we will meet again. Emulsions thicker than 100 μm are difficult to process and also suffer from loss of resolution if the beam does not pass normally through the emulsion. To obtain a resolution of 1 μm, the width of the projection of the ray in the emulsion must not exceed this and we note that in a 100 μm emulsion the ray must be within $\frac{1}{2}^o$ of the normal whereas with a 25 μm emulsion a 2° error can be tolerated. Thin emulsions give a greater statistical fluctuation in the number of developed grains per unit area and, of course, necessitate longer exposures.

Due to the dense packing of grains, it is difficult to obtain uniform development and in order to reduce the development rate so it is comparable with the diffusion rate of developer through the emulsion, development should be performed at about ice temperature. We carry out developing in the main body of a domestic refrigerator running as cold as possible. Prior to development, the emulsion must be softened by soaking. The developer, Kodak D19b, keeps reasonably well when concentrated but goes off rapidly when

diluted. The developer should therefore be changed <u>daily</u>. The stop bath
and fixer can, however, be used repeatedly. Plates should be washed in
cold, filtered tap-water for several hours, and then gently swabbed in running water with cotton wool to remove any remaining particles. Extreme care
must be taken in keeping the solutions clean if good results are to be
obtained. Solutions should be filtered before use.
Table 2.3 shows a processing schedule for Ilford emulsions.

TABLE 2.3. Processing times (in minutes) for Ilford L4 emulsions

Thickness	50 μm	25 μm
Soak in filtered deionized water	10	5
Develop (1:3 D19b to deionized water)	15-60	12-30
Stop (1% glacial acetic acid in deionized water)	10	5
Fix (300g sodium thiosulphate 30g sodium bisulphite in 1 litre deionized water)	60	30
Wash (filtered tap-water)	upwards of 120	upwards of 120

Plates should be dried under cover in air and not force dried. We find
it convenient to place the plates emulsion downwards on watch-glasses. The
emulsion can be protected and light scattering from the silver grains reduced by placing a drop of immersion oil on the topograph and covering with a
microscope cover slip. Care must be taken not to introduce air bubbles under
the cover slip. Lang (1970) recommends fixing the cover-slip in an air-
hardening medium such as "Eukitt".
Over a considerable density range, the Ilford emulsions are linear and it
is therefore possible to compensate for incorrect exposure in the development procedure. While it is better to have an overexposed, underdeveloped
plate rather than vice versa, underexposure can be compensated to a fair
degree, though the range over which this is possible is limited. One should
aim for a basic photographic density of unity from the perfect diffracting
crystal in the thin crystal situation and then the dark "direct" images give
densities up to 2 or 3. In thick crystal anomalous transmission topographs
a basic density of about 2 from the perfect crystal gives a good contrast
from the (white) dynamical images but still enables increases in intensity to
be observed. A general rule of thumb is to develop until a clear image is
seen on the back of the plate.
On development, the grains of the L4 emulsion swell to about 0.25 μm.
However, the granularity observed on the topograph when viewed under the
microscope is not a result of the grain of the film but is statistical 'shot
noise', arising from the statistical variation in the number of developed
grains per unit area. This granularity can be reduced by increased exposure,
but this may cause overexposure. Lang (1970) has considered the effect of
this noise on resolution in some detail.

When using hard radiations (MoKα or AgKα) the resolution will be limited also by the length of the photoelectron tracks in the emulsion which are about 2 μm long with these radiations. As we can see, many factors conspire to limit topographic resolution to between 1 and 2 μm).

2.7. ENLARGEMENT OF TOPOGRAPHS

There is little point in taking great care over the production and processing of the initial plate if the subsequent optical enlargement is poor. Each step in the making of a topograph is as critical as its predecessor. The most difficult thing to achieve in enlarging the topograph is low light scattering from the emulsion, and the collimation of the incident light is crucial. Topographs are generally viewed under transmitted light, but when plates are overexposed or overdeveloped, information otherwise lost can be reclaimed by viewing in reflected light (Mardix, 1974). Experience shows that the best topographs come from the best (and most expensive) microscopes. The Reichaert metallurgical microscope or the Zeiss "Ultraphot" microscope give excellent pictures and combine high quality with the low magnifications required for topography. Magnifications over 500× are almost never used, while 30× to 60× provides a very useful range of enlargement. Photography should be performed on the microscope, and topographs are conventionally printed as positives, i.e. as seen on the nuclear plate, excepting anomalous transmission topographs where the precedent set by Young is to print as negatives. For high resolution topographs it is not usually satisfactory to magnify topographs with a photographic enlarger. Light scattering reduces the contrast of dislocations, and at the necessary magnification the depth of field is small. In crystal growth studies, where the individual images of dislocations become of secondary importance to the overall defect configuration, this form of enlargement is often necessary. Such enlargements, between ×2 and ×10 are extremely difficult to perform on a microscope even when fitted with a macro-attachment.

2.8. RAPID HIGH RESOLUTION TOPOGRAPHY

By now the reader will probably have noticed that the conditions for high resolution and short exposure are mutually exclusive. As one of the main complaints about X-ray topography is the long exposure times, it is desirable to optimise the system to be as efficient as possible. The calculations presented here are straightforward but often overlooked.

Consider, as in Fig. 2.8, the intensity dI reaching the crystal due to a small area dX, dY of the X-ray source. This is

$$dI = P \ dX \ dY/HVD^2, \qquad\qquad (2.5)$$

where P is the total X-ray tube power, H the horizontal dimension, V the projected vertical dimension of the source, and D the source to specimen distance. In the vertical direction, the whole divergence of the beam is utilized, but in the horizontal direction only a fraction $\Delta\theta_B$ of the

reflecting range is diffracted. The total diffracted intensity becomes

$$I \propto (P/HVD^2) \int_0^V dX \int_0^{D\Delta\theta_B} dY = P\Delta\theta_B/HD. \qquad (2.6)$$

Assuming a linear response of the film, the exposure time t for a stationary topograph is

$$t = (C\Delta\theta_B P/HD)^{-1}, \qquad (2.7)$$

with C a constant of proportionality.
   For a traverse topograph

$$T = \{C\Delta\theta_B (P/HD)(M/R \cos \theta_B)\}^{-1}, \qquad (2.8)$$

where R is the length of traverse and M the slit width.

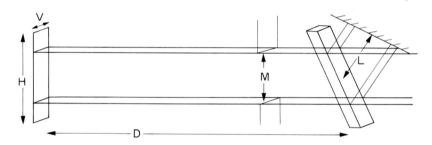

Fig. 2.8. Parameters influencing exposure time. In the diagram,
        the plate is just sufficiently far from the crystal to
        separate the direct and diffracted beams.

   In order to minimise the exposure time, we chose M = H, so the whole source is used. Then, neglecting the distance from the collimating slit to the specimen, we require for monochromaticity

$$H = M < D\theta_{\lambda_1 \lambda_2}, \qquad (2.9)$$

where $\Delta\theta_{\lambda_1 \lambda_2}$ is the angular separation between adjacent X-ray lines.
Substitution of (2.9) into (2.7) and (2.8) gives

$$t = (C\Delta\theta_B \Delta\theta_{\lambda_1 \lambda_2} P/H^2)^{-1} \qquad (2.10)$$

for the stationary topograph and

$$T = (C\Delta\theta_B \Delta\theta_{\lambda_1\lambda_2} P/HR \cos\theta_B)^{-1} \qquad (2.11)$$

for the traverse topograph.

These equations give the optimum conditions for a rapid high resolution topograph when only the horizontal resolution is considered. However we must also take the vertical resolution into account, and that is given by expression (2.2). In order to separate direct and diffracted beams,

$$L > M/\tan 2\theta_B. \qquad (2.12)$$

That is

$$\delta > V\Delta\theta_{\lambda_1\lambda_2}/\tan 2\theta_B. \qquad (2.13)$$

With (2.10), (2.11), and (2.13) we are in a position to optimize our topography and also to compare the merits of the transmission Berg-Barrett and Lang methods. We note, however, that (2.12) is not applicable to Lang's technique as it is physically impossible to place the plate closer than about 1 cm from the crystal. Under optimum conditions the vertical resolution of a Lang topograph will be

$$\delta_L = VL\Delta\theta_{\alpha_1\alpha_2}/H. \qquad (2.14)$$

The importance of always placing the plate as close as possible to the specimen is clearly demonstrated. Substitution of typical values (V = H and $\Delta\theta_{\alpha_1\alpha_2}$ = 2 x 10^{-4}) gives a resolution of 2 μm when L = 1 cm. As the value of $\Delta\theta_{\alpha_1\alpha_2}$ is a lower limit, corresponding to a Bragg angle of 6° we note that it is only with extreme care that the limits of resolution can be reached.

Vertical resolution can be improved by reducing the take off angle and it is worth noting that relatively rapid topography with a resolution of about 3 μm can be achieved from a 8 x 0.4 mm^2 source when the line is viewed at 2°. Comparative figures for the Phillips/Elliott sealed tube and the Elliott GX6 rotating anode generator running at 3000 rpm are given in table 2.4.

In purchasing a generator for topography the source dimensions and particularly the power per unit length of line (P/H) are most important. A survey conducted in 1969 showed that P/H varied with cost as

$$P/H \propto (cost)^{5/4}. \qquad (2.15)$$

Many workers have devoted considerable effort towards reducing exposure times and also eliminating the high precision moving parts encountered in the Lang camera. Probably the only significant improvement in speed without loss of resolution is the curved crystal technique of Wallace and Ward (1975). Unfortunately it is only applicable to thin plates of brittle materials. The relative speed of the Lang and transmission Berg-Barrett methods is a function of the source dimensions and in some cases it may be that Lang topography is not the most rapid method. Use of the Kβ lines is usually slower because of the lower intensity of the Kβ to the Kα_1 line, but use of the $K\alpha_1$ line with an oscillating Soller slit can result in an improvement.

TABLE 2.4. Comparative optimum speeds for Lang topography

	Source projected (mm^2)	Power (kW)	T (arbitrary units)
Sealed tube 8 x 0.4 mm^2 (taken off at 2°)	0.3 x 0.4	1.2	3
GX6 rotating anode tube 2 x 0.2 mm^2 (taken off at 6°)	0.2 x 0.2	2.0	1

TABLE 2.5. Comparative speeds of Lang and wide beam topographs (using Silicon 111 reflection)

Technique	Scan width (mm)	Power (kW)	Source (mm^2)	H (mm)	V (mm)	δ (μm)	P/H (kW/ mm)	Relative exposure time
Sealed tube Lang $CuK\alpha_1$ 2° take-off	8	1.2	8 x 0.4	0.4	0.3	4	3	10
Trans. B-B $CuK\beta$ 4° take-off	8	1.2	8 x 0.4	8	0.03	1.3	0.15	25
GX6 rotat- ing anode Lang $CuK\alpha_1$ 4° take-off	3	3.4	3 x 0.3	0.3	0.2	4	11.3	1
Trans. B-B $CuK\beta$ 4° take-off	3	3.4	3 x 0.3	3	0.02	1	1.1	1.3

2.9. DIRECT VIEWING OF X-RAY TOPOGRAPHS

The photographic method, while giving excellent resolution and contrast, is tedious and generally inapplicable to dynamic experiments in which the defect configuration varies on a time scale of seconds or minutes. This latter objection may be removed by the increased use of synchrotron radiation for topography (where exposure times are of the order of seconds on nuclear emulsions), but the desire to actually <u>see</u> what is going on is very strong amongst topographers. Several dynamic viewing systems have been developed and the early work is reviewed by Green (1971). Devices applied to topography fall into two categories – those employing direct conversion of X-rays to electrical signals and those using an X-ray to optical converter.

2.9.1. Direct Conversion

Without doubt, this has been the most successful of the direct viewing systems. As developed by Chikawa and colleagues (Chikawa and Fujimoto, 1968, 1972), the diffracted X-ray beam passes through a beryllium window onto a thin PbO photocathode of a sensitive vidicon TV tube (Fig. 2.9). The mode of operation differs slightly from that normally used. When the PbO target is made photoconducting by the X-rays, conduction of this charge takes place towards the beryllium faceplate which is maintained at a positive potential with respect to the target. This causes the target potential to rise towards that of the faceplate, and as the scanning beam passes on the next scan enough electrons are deposited to restore the target potential to its initial value. Capacitive coupling between target surface and face plate produces a voltage drop across an external load resistor, and on amplification this becomes the video signal. The sensitivity of the device is finally limited by the target thickness. A thin target absorbs only a small fraction of the X-rays, and in thick targets only a thin layer on the X-ray entrance surface side becomes photoconducting.

Initially, Chikawa obtained a resolution of 30 μm, but recently, by raising bias voltage, a resolution of 15 μm has been achieved. Individual dislocations in silicon are easily resolvable (Fig. 2.9) and the topographs of moving dislocations in silicon (Chikawa, Fujimoto, and Abe, 1972) and of the growth of crystals *in situ* are truly impressive. (See Chapter 7).

A similar system developed by Rozgonyi, Haszko, and Statile, (1970) using a silicon diode array camera gave an impressive 15 μm resolution when used for reflection Berg-Barrett topography with CrKα radiation from a conventional generator. However, with hard radiation and transmission topography, a very much higher X-ray flux is required to give similar resolution and also to get above the dark current in the tube. As it is illustrative of the problems of direct display we will briefly consider the criteria for good resolution using Chikawa's system as an example.

Consider a flux I/unit area/unit time of X-ray photons incident on a detector of efficiency η. The number of photons N detected in an integration time τ in a square picture element of side ε is then

$$N = I\eta\varepsilon^2\tau. \tag{2.16}$$

Suppose two elements in the topograph have an intensity difference ΔI, then the difference in signal ΔN is

$$\Delta N = \Delta I\eta\varepsilon^2\tau. \tag{2.17}$$

When written in terms of the visibility V, defined as $V = \Delta I/(2I + \Delta I)$, this becomes

$$\Delta N = \eta \epsilon^2 \tau 2IV/(1 - V). \qquad (2.18)$$

Now the r.m.s. noise on the signal is

$$\left| N + (N + \Delta N) \right|^{\frac{1}{2}} = (2N + \Delta N)^{\frac{1}{2}}. \qquad (2.19)$$

The signal-to-noise ratio R is then given by

$$R = \Delta N(2N + \Delta N)^{-\frac{1}{2}} = (V\Delta N)^{\frac{1}{2}}, \qquad (2.20)$$

and substitution into (2.18) gives

$$\epsilon = \frac{R}{V}\left(\frac{1 - V}{2\eta\tau I}\right)^{\frac{1}{2}} \qquad (2.21)$$

This fundamentally important equation defines the resolution of the system ϵ and we note that several factors are crucial to good resolution.

1. High detection efficiency η

We have already highlighted the problem. A thin target is inefficient leading to poor resolution but a thick one gives a poor resolution in any case. A compromise on target thickness has to be reached.

2. Long integration time τ

This will generally be limited by the electronics, but for dynamic experiments should not be more than a few seconds.

3. High X-ray intensity I

Chikawa used a Rigaku Denki rotating anode generator delivering 30 kW on a focal spot of 0.5×10 mm^2, which is the maximum intensity available from present conventional generators. The recent application of synchrotron radiation to topography (Tuomi, Naukkarinen and Rabe, 1974; Hart, 1975) opens great possibilities for high resolution operation of a direct viewing system but has yet to be proven. A possible source of trouble is the rate of degradation of the PbO target under irradiation.

4. Visibility V

This is not a parameter to be varied in the same sense as the others, but we note that the higher the visibility the better the resolution. Thus, gross defects will be easier to image than individual dislocations. The best dislocation visibility occurs under low absorption conditions, and these should be used whenever possible.

The main disadvantage with the TV system is that it is expensive. Many applications require a rapidly produced image of low resolution for setting

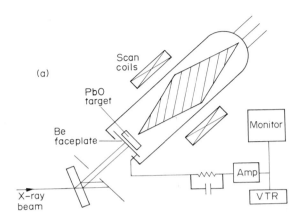

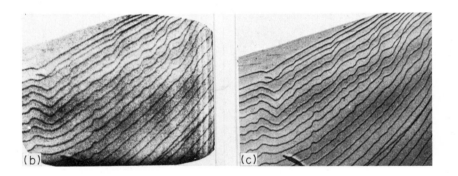

Fig. 2.9. (a) Schematic diagram of Chikawa's direct topographic
         imaging system. (b) Video topograph taken from the out-
         put monitor of dislocations in silicon. Field width 9 mm.
         (c) The same image recorded on nuclear emulsion film
         (courtesy J. Chikawa).

up purposes and one cheap solution is to use dental film. Channel plates
have been used successfully to record topographs (Parpia and Tanner, 1971)
and are inexpensive.

    Unfortunately, while channel plates are sensitive to long wavelength
X-rays and relatively sensitive to hard X-rays such as used in medicine, a
minimum in efficiency occurs in the crystallographic wavelengths. Using the
lead glass matrix as an absorber, one can calculate the probability of a
photoelectron finding its way into a channel. One finds (Tanner, 1971) that
the efficiency

$$\eta = F\mu R_p \cot \emptyset/2\pi - F\mu^2 R_p^2 \cot \emptyset \, \text{cosec} \, \emptyset/18, \qquad (2.22)$$

where μ is the X-ray absorption coefficient, R_p is the effective range of the ejected photoelectrons, F is the fractional area of the faceplate consisting of open channel and $\emptyset$ is the angle of the beam to the plate normal. Substitution of appropriate numbers for an angle of incidence of 5° yields an efficiency of 2% for CuKα radiation. This is about what is found experimentally, and for high detection efficiencies optical conversion is desirable.

2.9.2. X-ray to Optical Conversion

Probably the most successful is the system of Lang (Lang and Reifsnider, 1969; Lang, 1971) which has been demonstrated to image individual dislocations if in a somewhat ideal situation. The device, similar to that developed earlier by Meieran, Landre, and O'Hara (1969) uses a thin fluorescent screen to convert the X-ray image to an optical one and then intensifies this by a high-gain image intensifier. The image formed on the output screen of the intensifier may be viewed directly or with a conventional vidicon CCTV camera. As in the direct conversion systems, the resolution is statistically limited and great care must be taken to ensure a minimum of light loss between components. Fibre optic coupling of the screen to intensifier is ideal in this respect, but one then may be limited by the intrinsic resolution of the intensifier. Lang has attempted to optimize the situation by using a single-stage high-resolution intensifier as a "preamplifier" prior to the lower resolution high-gain intensifier. The preamplifier can be coupled to the screen fibre optically and the more intense output magnified onto the second intensifier.

The ultimate limit to the performance of optical conversion systems, assuming perfect optical image intensifiers, comes in the fluorescent screen. For good resolution we require a thick screen (for maximum conversion efficiency), but a thick screen gives poor spatial resolution due to spreading of the light through the screen. Lang compromised on a CsI screen polished to 100 μm, but a promising recent development is the evaporated CsI screens now being used for medical X-ray intensification. The vapour-grown CsI crystals grow as needles normal to the substrate and the grain boundaries act as diffuse scatterers which effectively throws the light in the forward direction. There appears to be no limit to the minimum thickness of the screen.

Lang has produced an impressive ciné film of moving moiré fringes and dislocations in diamond taken with a conventional fine focus tube. Hashizume *et al.* (1971) have demonstrated that when high spatial resolution is not required, as in some double crystal applications, a fluorescent screen backed by a sensitive image orthicon TV camera can be very successful.

2.10. DOUBLE CRYSTAL TOPOGRAPHY

Although at first sight the single crystal topographic techniques appear to be sensitive to strains of the order of 1 part in 10^5, in practice they are found to be very insensitive to slow variations of lattice parameter. This arises because the large angular divergence of the X-ray beam results in the diffracted intensity being equivalent to the integrated intensity. Until the lattice parameter variations become of the order of one part in 10^3, the intensity in the image is constant. However, use of two crystals enables such long range variations to be detected.

The double crystal technique (Bond and Andrus, 1952; Bonse and Kappler,

1958), as the name implies, utilizes two successive Bragg reflections.
Figure 2.10 shows a schematic diagram of the (+−) parallel setting where the
Bragg planes of the two crystals are parallel and of equal spacing. In this
geometry, the double crystal technique is extremely sensitive to lattice
distortion or misorientation and the rocking curve obtained when one of the
crystals is rotated is very narrow. In fact, it is just the convolution of
the two perfect crystal reflecting curves and has a width about 1.4 times the
perfect crystal reflecting range. We can understand how this arises by
reference to the Du Mond diagram in Fig. 2.11a.

The Du Mond diagram shows the angles at which Bragg reflection occurs for
various wavelengths from a given set of lattice planes, and is essentially
a graphical representation of Bragg's law. Where the curves for the two
crystals overlap, simultaneous diffraction occurs. As the Bragg planes are
parallel and equispaced, the curves overlap for all values of (λ, θ) satisfying reflection, and in the exploded view we see the shaded area corresponding
to the perfect crystal reflecting range convoluted with that of the second
crystal. We note that the (+−) parallel setting is non-dispersive, i.e. all
wavelengths are diffracted at the same relative setting of the two crystals.
If there is a local misorientation in the second crystal, that local part of
the crystal will display a Du Mond diagram as shown in Fig. 2.11b. The
curve is misplaced and no region of overlap occurs for any wavelength and
local loss of intensity results. Similarly, distortion yields a Du Mond
diagram as in Fig. 2.11c and again no diffraction occurs. The angular range
over which simultaneous diffraction occurs is thus very narrow and may be as
low as 0.1 second of arc.

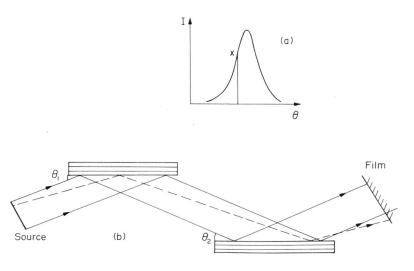

Fig. 2.10. The double crystal technique. (a) Rocking curve showing
        the position of the crystal on the flank of the curve
        used for high strain sensitivity. (b) (+−) parallel
        setting, illustrating the non-dispersive nature of the
        system. In this arrangement, the lattice plane spacing
        of the two crystals are equal.

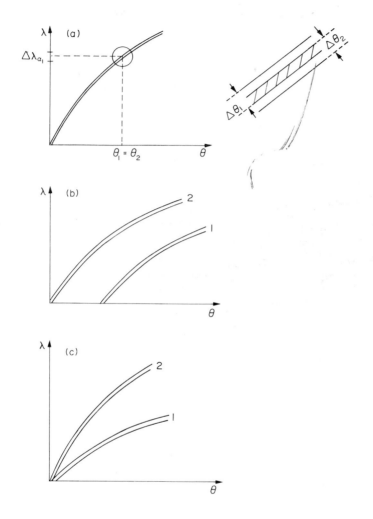

Fig. 2.11. (a) Du Mond diagram corresponding to two perfect crystals in
the (+-) parallel setting. (b) Du Mond diagram corresponding to local rotation of the lattice in the second
crystal. (c) Du Mond diagram corresponding to local
dilation in the second crystal.

   In normal operation the second crystal is set on the steep flank of the
rocking curve, say at point X in Fig. 2.10. Here, to a good approximation,
the slope of the rocking curve is linear and the relative change in intensity
can be related to the lattice distortion by a simple geometrical relation
(Bonse, 1962). Explicitly,

$$\Delta I/I = k(\tan \theta \, \Delta d/d \pm \underline{n}_g . \underline{n}_t \Delta \theta), \qquad\qquad (2.23)$$

where k is the slope of the rocking curve at X, $\underline{n}_g$ is a unit vector normal to the incidence plane, and $\underline{n}_t$ is a unit vector parallel to the tilt axis.

This simple theory has been found to give a good description of most of the contrast found in double crystal topographs. Using short wavelength radiation and high order reflections, Hart (1968) has obtained sensitivities of $k = 10^6$ in silicon. A change of intensity of 1% in the topograph then corresponds to a fractional change in the lattice parameter of one part in 10^8. In his studies of low oxygen float grown silicon (LOPEX) Hart (1968) found lattice parameter fluctuations of about one part in 10^7.

The increased sensitivity of the (+−) parallel setting over the Lang technique is apparant in Figs. 2.18a and b. Growth striations, clearly visible in the double crystal topograph, are totally invisible in a Lang topograph.

For high sensitivity, it is essential to use the same material for both specimen and reference crystal and also to ensure that the twist between the Bragg planes of the two crystals is small. In such topographs, the images of dislocations appear very broad as high contrast is obtained from the small strains present far from the dislocation core. High sensitivity double crystal topography is therefore not suited to crystals with high dislocation densities. Even with dislocation-free crystals, one cannot consider the reference crystal as being "perfect" and the image must be considered as arising from defects in both crystals. Fortunately, as the first crystal is a fair distance from the photographic plate, images of defects in this crystal appear blurred and are not usually troublesome. In his double crystal topographs of LOPEX silicon, Hart observed two sets of growth striations which he distinguished by a $90°$ rotation of one of the crystals.

When the specimen crystal has a lattice parameter variation large compared with the angular sensitivity of the (+−) double crystal arrangement, only part of the crystal diffracts for any point on the rocking curve. The result is a contour of misorientation on the plate. By stepping the specimen crystal across the rocking curve, many such contours may be recorded on one plate. Some of these topographs are extremely attractive visually, ranging from simple "Zebra stripes" to contorted ring patterns.

Although containing no wavelength dispersion, the (+−) parallel setting is angular dispersive and thus images of defects produced by different wavelengths are not formed at the same point on the photographic plate. This is usually overcome by introducing some dispersion by use of different materials for the reference and sample crystals. As in the (++) setting discussed below, only a range of wavelengths diffracts at any setting, and the angular sensitivity is correspondingly reduced. The (+−) setting with different crystals is obviously a very flexible mode of operation. The reference crystal can be chosen to be highly perfect and with Bragg plane spacing close to that of the sample crystal. Lattice parameter variations of typically one part in 10^5 can be revealed in this way. Okazaki and Kawaminami (1973) have shown that by use of white radiation at an $89°$ Bragg angle they were able to measure the change in lattice parameter of $KNiF_3$ as a function of temperature to one part in 10^5. The application of double crystal diffractometry to the measurement of the difference in lattice parameter of thin films and their substrates is mentioned in Chapter 4.

In the (++) setting, we have equal Bragg plane spacings but non parallel Bragg planes. The corresponding Du Mond diagram is shown in Fig. 2.12 and we see that for any relative orientation of the crystal a narrow band of

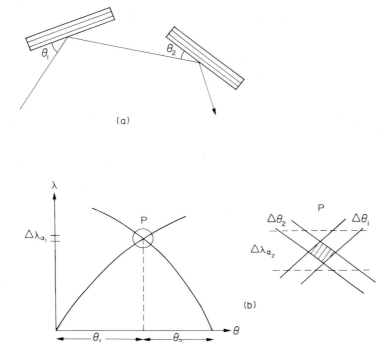

Fig. 2.12. (a) (++) double crystal setting. (b) Corresponding
        Du Mond diagram.

wavelengths only will be diffracted. The system is thus dispersive, and the
range of wavelengths $\Delta\lambda$ diffracted is much smaller than the natural line
width of the characteristic lines. When the crystal is locally misoriented,
the intersection point P moves along the curves and thus within the range of
the characteristic linewidth little change in contrast is observed.
Relatively large misorientations can now be tolerated without loss of intensity and the angular sensitivity is comparable with the Lang technique.
Double crystal topography in the (++) setting has been studied and developed
by Kohra and co-workers (Kohra, Hashizume, and Yoshimura, 1970; Nakayama
et al., 1973). Defect contrast is, however, complex and has not yet been
fully investigated and the setting has not spread into widespread use.
    An important feature of double crystal topography is that asymmetric
reflections can be used to broaden the beam to cover the whole of the specimen area (Fig. 2.13). This widening of the beam also leads to a narrowing
of the reflecting range, as may be seen from (1.63A). Kohra has also
explained this in terms of an effective lens action (Fig. 2.14) in which the
source is effectively moved further from the crystal and hence narrows the
angle of divergence. Extensive studies have been made on the use of asymmetric reflections in collimators (Kikuta and Kohra, 1970; Kikuta, 1971).
Beams have been produced with sufficiently small divergence that structure

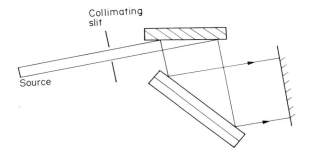

Fig. 2.13. (+-) parallel setting with asymmetrically cut crystals.

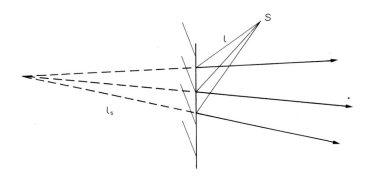

Fig. 2.14. The effective lens action in asymmetrically cut crystals.

factor measurements can be made from the intrinsic reflecting curve width.
Such a technique has been employed on germanium (Matsushita and Kohra, 1974).
    Perhaps a reflection of the relative under-use of double crystal techniques
is the fact that a double crystal diffractometer as such is not marketed,
although a second axis can be bolted onto a standard APEX diffractometer as
required. To perform high sensitivity work, however, considerable care
should be taken in the construction of the camera. This should be cast in
one piece with water cooling of the base and particular attention paid to the
alignment of the axes, bearings being pre-loaded and all parts machined to
high precision. Crystals can be driven, via micrometers, by stepping motors
and piezoelectric control used for fine adjustment. The group at Edinburgh
directed by Milne will supply custom built cameras.
    One slightly unusual double crystal camera which is available commercially
is the "OMD" camera originally developed by Kohra and Takano (1968). In this
system, a divergent monochromatic beam is produced by reflection from a
curved reference crystal. A slit is placed on the Rowland circle to allow

only the passage of the $K\alpha_1$ line onto the specimen crystal. The technique
is faster and less sensitive to orientation contrast than Lang's technique
and is particularly useful for crystals containing sub-grains.

High precision measurements of relative lattice parameters may be made
using multiple Bragg reflections in a double crystal arrangement. Hart
(1969) showed that with two crystals cut from the same block of silicon and
left standing on it a precision of one part in 10^9 was achievable. In the
original design, two X-ray sources were used but recently experiments have
been designed where only one source is necessary. Very recently Baker
et al. (1976) have measured the lattice parameter of gallium arsenide relative to a highly perfect silicon reference crystal to 2 parts per million
using such a technique.

2.11. X-RAY MOIRÉ TOPOGRAPHY AND INTERFEROMETRY

One of the important points to emerge from Chapter 1 was that in regions
where refracted and transmitted beams overlap, the wavefield is spatially
modulated with a periodicity corresponding to the Bragg plane spacing. If we
now place a second crystal close to the exit surface of the first crystal
(Fig. 2.15) within the standing wavefield moiré effects can be observed

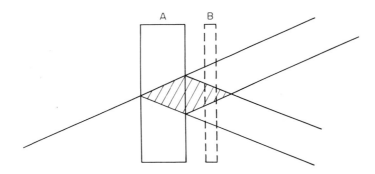

Fig. 2.15. The moiré sandwich arrangement where the second crystal B
          is placed within the standing wavefield emerging from the
          first crystal A.

between the modulated standing wavefield and the grid of the second crystal
lattice planes. The X-ray moiré sandwich was developed by Bradler and Lang
(1968) following observation of X-ray moiré fringes in quartz (Lang and
Muiscov, 1967) and CdS (Chikawa, 1967) The moiré fringe spacing D is given
by

$$D^{-1} = \underline{g}_1 - \underline{g}_2 \tag{2.24}$$

where $\underline{g}_1$ and $\underline{g}_2$ are the reciprocal lattice vectors for the two crystals.

Both rotation and dilation moiré fringes have been observed and behave
similarly to optical moiré fringes. An elegant demonstration of the dis-
continuity of the fringes at dislocations has been given by Lang (1968) and
is shown in Fig. 2.16.

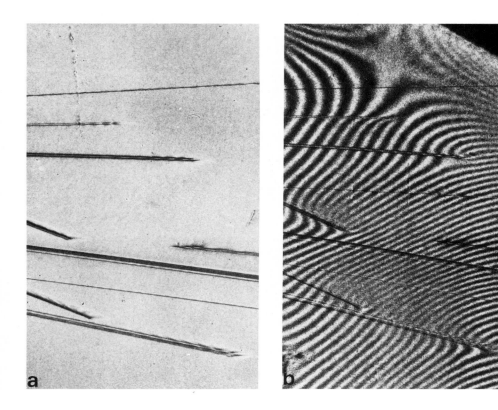

Fig. 2.16. (a) Projection topograph of dislocations in synthetic
          quartz. (b) Corresponding moiré topograph when the
          crystal is incorporated into a sandwich with another,
          perfect, quartz crystal. Note the discontinuity of the
          fringes at the dislocation lines which outcrop at the
          inner face of the sandwich. There are g.b fringes at
          each line (courtesy A.R. Lang).

The sandwich is extremely hard to set up and adjust. Bradler and Lang
devised a jig for rotating the crystals to bring the reciprocal lattice
points into coincidence but there is still the problem of holding the
crystals stable relative to each other on the Ångstrom scale during the
course of the exposure. This was solved by sticking the crystals together

though trouble often occurred during the setting of the adhesive. The most
common way of using moiré topography is in the X-ray interferometer.
Interferometers (Bonse and Hart, 1965; Hart, 1971) are usually monolithic
and cut from a single block of silicon using diamond loaded milling tools.
The refracted and diffracted beams are recombined a considerable distance
from the reference crystal (beam splitter) by reflection from two mirror
crystals (Fig. 2.17). An analyser crystal intersects the standing wavefield

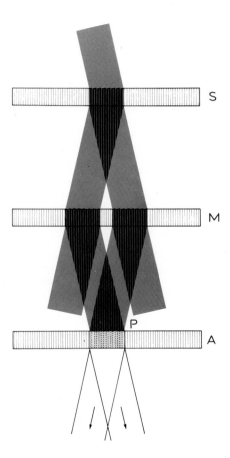

Fig. 2.17. The principle of the X-ray interferometer (courtesy
          A.D. Milne).

produced by the recombined beams. In a monolithic device, lattice continuity is preserved between beam splitter, mirrors and analyser, so in a perfect interferometer the intensity across a diffracted beam emerging from the
analyser would be uniform. However, a small twist imparted to the analyser
would lead to a rotational moiré pattern and, similarly, dislocations or

other lattice defects generate moiré patterns also.

An example of the moiré pattern generated by a highly perfect silicon crystal is given in Fig. 2.18c. Upon oxidation, the moiré fringes contract as the lattice distortion becomes large (Fig. 2.18d). Note that while the interferometric technique is capable of detecting quite small lattice parameter changes – a relative distortion of one part in 10^8 will yield moiré fringes of spacing about one centimetre – it does not have a high spatial resolution. For such resolution and sensitivity the double crystal method should be used. An excellent review of interferometry has been given in the review article by Hart (1971).

Many novel applications have been devised for interferometers. One of the most important is to obtain an absolute value for the silicon lattice parameter. From this secondary standard of length it should be possible to obtain a value of Avogadro's number to a much higher precision than is currently possible. The essence of the experiment is to count the number of moiré fringes, and hence the number of lattice planes, passing when the analyser crystal is translated by a given distance. This distance is measured by making the analyser part of an optical interferometer. Devices are either monolithic (Curtis *et al.*, 1971) or polylithic (Bonse and te Kaat, 1968; Deslattes 1969). Current work is in progress to establish the accuracy and reliability of the measurements.

The X-ray interferometer may also be used as a genuine interferometer to measure X-ray refractive indices (Bonse and Hart, 1965). An interferometer is known as an ideal interferometer when the path from splitter to mirror equals that from mirror to analyser for both O and G beams. When this occurs we have, for thick wafers, an intensity maximum at the analyser. Introduction of a phase object (e.g. a plastic lens) can produce a varying phase shift across one of the beams. On recombination at the analyser a pattern is seen in the emergent beams corresponding to contours of equal thickness in the lens. Measurement of the fringe spacing gives the refractive index directly.

A recent new development is the neutron interferometer, and Bonse and co-workers have used this in a similar fashion to measure neutron scattering lengths (Rauch, Treimer, and Bonse, 1974).

2.12. SYNCHROTRON TOPOGRAPHY

Probably the two most frequent complaints about X-ray topography are that exposure times are long, and that very critical adjustments are necessary when setting up the crystal. It therefore came as a pleasant surprise when just over a year ago, Tuomi *et al.* (1974) showed that topographs could be obtained in the order of seconds simply by placing a crystal in a beam of synchrotron radiation and recording images of the diffracted beams. The technique used is essentially the transmission equivalent of the Schulz method, originally developed by Guinier and Tennevin (1949).

An electron constrained to a circular orbit emits electromagnetic radiation, which in the highly relativistic limit is seen in the laboratory frame as a narrow cone tangential to the electron orbit. At 5 GeV the photon flux peaks around 1 Å and the angular divergence is about 10^{-4} radians. On the electron synchrotron NINA at Daresbury, U.K. the effective source size is 0.5 mm square and the experimental area 50 m from the synchrotron ring. The radiation is plane polarized in the plane of orbit and is a continuous spectrum extending through the visible and ultraviolet into the X-ray region.

A little reflection on the above specification will reveal that each

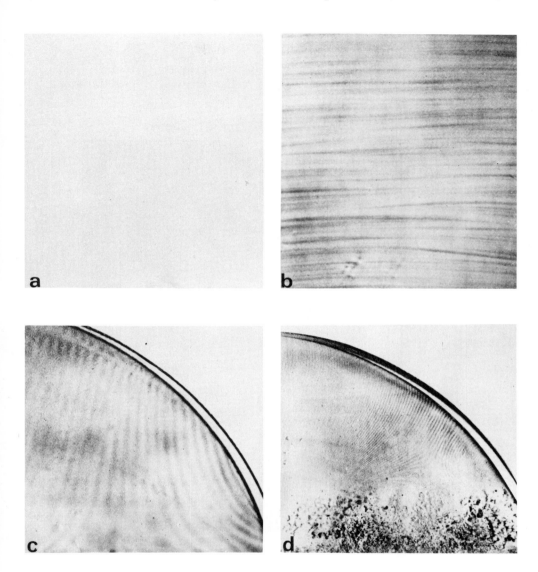

Fig. 2.18. Comparison of Lang, double crystal and interferometric
techniques. (a) Lang topograph in transmission, 220
reflection, MoKα_1 radiation. (b) Double crystal topograph. Symmetrical reflection 880 in the (+−) parallel
setting. (c) Interferometric pattern from an as-grown
crystal. Wafer 500 μm thick. 220 reflection, MoKα
radiation. (d) Same interferometer as in (c) after
oxidation for 2 hours. The crystal used in all topographs was float zoned silicon (courtesy A.D. Milne)

crystal lattice plane will select a particular wavelength for diffraction
and we obtain what is essentially a Laue pattern, but with each "spot" now
being an image of the crystal. Crystal defects are seen within each topograph, Fig. 2.19. Due to the relatively wide beam (up to 30 mm diameter on

Fig. 2.19. Synchrotron topographs of a crystal of $RFeO_3$ taken with
the continuous radiation from NINA at Daresbury.
Exposure 30 sec on 10 μm thick Ilford L4 emulsion.

NINA) scanning of the crystal across the beam is not required and it is
necessary only to adjust the crystal orientation to approximately a degree
in order to select the approximate wavelength required in any particular
diffracted beam. Although many images appear at once, two beam conditions
are found as each image is formed with different wavelength X-rays. Burgers
vector analysis can, in favourable conditions, be performed with one

exposure. The photon flux is extremely high compared with conventional X-ray
sources and one finds exposure times of the order of seconds when using
Ilford L4 emulsions.

A further important feature is that for a geometrical resolution of $1\mu m$ on
NINA, the photographic plate may be placed 10 cm from the crystal. Hart
(1975) has demonstrated that provided care is taken to ensure that the plate
is normal to the diffracted beam, a resolution comparable with that obtained
using Lang's technique can be achieved.

The high flux, the redundancy of a scanning mechanism and the ability to
place the photographic plate a long way from the specimen enables many new
topographic experiments to be performed. A large amount of associated
equipment can now be placed around the sample in a manner previously
impossible. We have demonstrated that topographs of antiferromagnetic domain walls can be taken at low temperatures in fields up to 1.3 Tesla
(Tanner et al., 1976). Step by step observations of wall motion can be made
using nuclear emulsion plates as exposure times are so short. Allowing for
the time taken to operate the interlocks on the beam area, we are able to
take single exposures at a rate of one every 5 minutes. When we mounted six
plates on a motor driven wheel inside a lead lined light tight box, the time
per exposure was reduced below a minute.

The contrast of defects is somewhat more complicated than when using the
Lang technique. Firstly, in experiments where the specimen-plate distance
is large there is a mixture of extinction and orientation contrast which can
be quite difficult to disentangle. Secondly, the extinction contrast is
complicated by the presence of more than one order of diffraction within
each image. The details of the contrast have still to be studied either
experimentally or theoretically. However, there can be no doubt that the
use of synchrotron radiation will have far reaching effects on the future of
X-ray topography.

References

Angillelo, J. and Wood, C. (1971) *Rev. Sci. Inst.* **42**, 1709.
Aristov, V. V. and Shulakov, E. V. (1975) *J. Appl. Cryst.* **8**, 445.
Austermann, S. B. and Newkirk, J. B. (1967) *Adv. X-ray Analysis* (Plenum)
        10, 134.
Baker, J. F. C., Hart, M., Halliwell, M. A. G. and Heckingbottom, R.
        (1975) *Solid-State Electronics* **19**, 337.
Barrett, C. S. (1945) *Trans. AIME* **161**, 15.
Barth, H. and Hosemann, R. (1958) *Z. Naturforschung* **13a**, 792.
Berg, W. F. (1931) *Naturwissenschaften* **19**, 391.
Bond, W. L. and Andrus, J. (1952) *Am. Mineralogist* 37, 622.
Bonse, U. (1962) In *Direct Observation of Imperfections in Crystals*
        (ed. Newkirk and Wernick), p. 431, Wiley, New York.
Bonse, U. and Hart, M. (1965) *Appl. Phys. Lett.* **7**, 99.
Bonse, U. and Hart, M. (1965) *Z. Physik* **188**, 154.
Bonse, U., Hart, M., and Newkirk, J. B. (1967) *Adv. X-ray Analysis* (Plenum)
        10, 1.
Bonse, U. and te Kaat, E. (1968) *Z. Physik* **214**, 16.
Bonse, U. and Kappler, E. (1958) *Z. Naturforschung* **13a**, 348.
Bradler, J. and Lang, A. R. (1968) *Acta Cryst.* **A24**, 246.
Chikawa, J-I. (1967) *J. Phys. Chem. Solids* **28** suppl. 1, 817.
Chikawa, J-I. and Fujimoto, I. (1968) *Appl. Phys. Lett.* **13**, 387.
Chikawa, J-I., Fujimoto, I., and Abe, T. (1972) *Appl. Phys. Lett.* **21**, 295.

Curtis, I., Morgan, I., Hart, M., and Milne, A. D. (1971) In *Precise Measurements and Fundamental Constants* (ed. Landeberg and Taylor), p. 285, NBS Washington.

Deslattes, R. D. (1969) *Appl. Phys. Lett.* <u>15</u>, 386.

Dionne, G. (1967) *J. Appl. Phys.* <u>38</u>, 4094.

Gerward, L. and Lindegaard Anderson,A. (1974) *Phys. Stat. Sol. (a)* <u>23</u>, 537.

Green, R. E. Jr. (1971) *Adv. X-ray Analysis* (Plenum) <u>14</u>, 311.

Guinier, A. and Tennevin, J. (1949) *Acta Cryst.* <u>2</u>, 133.

Hart, M. (1968) *Science Progress* (Oxford) <u>56</u>, 429.

Hart, M. (1969) *Proc. Roy. Soc.* <u>A309</u>, 281.

Hart, M. (1971) *Rept. Prog. Phys.* <u>34</u>, 435.

Hart, M. (1975) *J. Appl. Cryst.* <u>8</u>, 436.

Haruta, K. (1965) *J. Appl. Phys.* <u>36</u>, 1789.

Hashizume, H., Kohra, K., Yamaguchi, T., and Kinoshita, K. (1971) *Appl. Phys. Lett.* <u>18</u>, 213.

Hosoya, S. (1968) *Jap. J. Appl. Phys.* <u>7</u>, 1.

Isherwood, B. J. and Wallace, C. A. (1974) *Phys. in Technology* <u>5</u>, 244.

Kikuta, K. (1971) *J. Phys. Soc. Japan* <u>30</u>, 222.

Kikuta, K. and Kohra, K. (1970) *J. Phys. Soc. Japan* <u>29</u>, 1322.

Kohra, K., Hashizume, H., and Yoshimura, J. (1970) *Jap. J. Appl. Phys.* <u>9</u>, 1029.

Kohra, K. and Takano, Y. (1968) *Jap. J. Appl. Phys.* <u>7</u>, 982.

Kuriyama, M. and McManus, G. M. (1968) *Phys. Stat. Sol.* <u>25</u>, 667.

Lang, A. R. (1958) *J. Appl. Phys.* <u>29</u> 597.

Lang, A. R. (1959) *Acta. Cryst.* <u>12</u>, 249.

Lang, A. R. (1963) *Brit. J. Appl. Phys.* <u>14</u>, 904.

Lang, A. R. (1968) *Nature* <u>220</u>, 652.

Lang, A. R. (1970) In *Modern Diffraction and Imaging Techniques* (ed. Amelinckx et al.), p. 407, North-Holland.

Lang, A. R. (1971) *J. Phys. E (Sci. Inst)* <u>4</u>, 921.

Lang, A. R. and Muiscov, V. F. (1967) *Appl. Phys. Lett.* <u>7</u>, 214.

Lang, A. R. and Reifsnider, K. (1969) *Appl. Phys. Lett.* <u>15</u>, 258.

Liang, S. J. and Pope, D. P. (1973) *Rev. Sci. Inst.* <u>44</u>, 956.

Mardix, S. (1974) *J. Appl. Phys.* <u>45</u>, 5103.

Matsushita, T. and Kohra, K. (1974) *Phys. Stat. Sol. (a)* <u>24</u>, 531.

Meieran, E. S., Landre, J. K., and O'Hara, S. (1969) *Appl. Phys. Lett.* <u>14</u>, 368.

Mellaert, L. J. van, and Schwuttke, G. H. (1970) *Phys. Stat. Sol. (a)* <u>3</u>, 687.

Nakayama, K., Hashizume,H., Miyoshi, A., Kikuta, K., and Kohra, K. (1973) *Z. Naturforschung* <u>28a</u>, 632.

Parpia, D. Y. and Tanner, B. K. (1971) *Phys. Stat. Sol. (a)* <u>6</u>, 689.

Okazaki, A. and Kawaminami, M. (1973) *Jap. J. Appl. Phys.* <u>12</u>, 783.

Oki, S. and Futagami, K. (1969) *Jap. J. Appl. Phys.* <u>8</u>, 1574.

Rauch, H., Treimer, W., and Bonse, U. (1974) *Phys. Lett.* <u>47A</u>, 369.

Roessler, B. (1967) *Phys. Stat. Sol.* <u>20</u>, 713.

Rozgonyi, G. A., Hasko, S. E., and Statile, J. L. (1970) *Appl. Phys. Lett.* <u>16</u>, 443.

Schulz, L. G. (1954) *Trans. AIME* <u>200</u>, 1082.

Schwuttke, G. H. (1965) *J. Appl. Phys.* <u>36</u>, 2712.

Tanner, B. K. (1971) *D. Phil. Thesis*, Oxford University.

Tanner, B. K., Safa, M., Midgley, D., and Bordas, J. (1976) *J. Magnetism and Magnetic Materials*.

Tuomi, T., Naukkarinen, K., and Rabe, P. (1974) *Phys. Stat. Sol. (a)* <u>25</u>, 93.

Turner, A. P., Vreeland, T., and Pope, D. P. (1968) *Acta Cryst.* <u>A244</u>, 52.
Wallace, C. A. and Ward, R. C. C. (1975) *J Appl. Cryst.* <u>8</u>, 281.
Wu, C. Cm. and Armstrong, R. W. (1975) *J. Appl. Cryst.* <u>8</u>, 29.

Appendix

Pioneer topographic experiment
Ramachandran, G. N. (1944) *Proc. Indian Acad. Sci.* <u>A19</u>, 280.

Berg-Barrett (reflection)
Merlini, A. and Guinier, A. (1957) *Bull. Soc. Fr. Min. Crist.* <u>80</u>, 147.
Newkirk, J. B. (1958) *Phys. Rev.* <u>110</u>, 1465.
Newkirk, J. B. (1959) *Trans AIME* <u>215</u>, 483.
Schiller, C. (1969) *J. Appl. Cryst.* <u>2</u>, 223.
Berg-Barrett (transmission)
Ando, M. and Hosoya, S. (1970) *J. Appl. Cryst.* <u>4</u>, 146.
Lindgaard Anderson, A. (1965) *Rev. Sci. Inst.* <u>36</u>, 1888.
Lindgaard Anderson, A. (1972) *Rev. Sci. Inst.* <u>39</u>, 774.
Gerold, V. and Meier, F. (1959) *Z. Physik* <u>155</u>, 387.
Tanner, B. K. and Humphreys, C. J. (1970) *J. Phys. D. (Appl. Phys.)* <u>3</u>, 1144.

Lang (reflection)
Lang, A. R. (1957) *Acta Cryst.* <u>10</u>, 839.
Yoshimatsu, M., Shibata, A. and Kohra, K. (1966) *Adv. X-ray Analysis*
          (Plenum) <u>9</u>, 14.
Lang (with chromatic aberration correction)
Chikawa, J-I., Fujimoto, I., and Asaedo, Y. (1971) *J. Appl. Phys.* <u>42</u>, 4731.
Stage design
Argemi, R., G'Sell, C., and Baudelet, B. (1971) *Rev. Sci. Inst.* <u>42</u>, 1711.
Middleton, R. M. and Caslavsky, J. L. (1972) *Rev. Sci. Inst.* <u>43</u>, 1662.
Zani, A. and Servidori, M. (1974) *J. Phys. E. (Sci. Inst.)* <u>7</u>, 592.
Hot stage
Kume, S. and Kato, N. (1974) *J. Appl. Cryst.* <u>7</u>, 427.
Limited Projection
Caslavsky, J. L. (1970) *Rev. Sci. Inst.* <u>41</u>, 517.

Double crystal
Bonse, U. and Hart, M. (1965) *Appl. Phys. Lett.* <u>7</u>, 238.
Hart, M. and Lang, A. R. (1965) *Acta Cryst.* <u>19</u>, 73.

Interferometry
Bonse, U. and Hart, M. (1966) *Z. Physik* <u>190</u>, 455.
Bonse, U. and Hart, M. (1966) *Z. Physik* <u>194</u>, 1.
Bonse, U. and Hart, M. (1968) *Acta Cryst.* <u>A24</u>, 240.
Becker, P. and Bonse, U. (1974) *J. Appl. Cryst.* <u>7</u>, 593.
Hart, M. (1968) *J. Phys. D. (Appl. Phys.)* <u>1</u>, 1405.
Hart, M. (1972) *Phil. Mag.* <u>26</u>, 821.
Hart, M. and Milne, A. D. (1969) *J. Phys. E. (Sci. Inst.)* <u>2</u>, 646.
Hart, M. and Bonse, U. (1970) *Physics Today* <u>23</u>, 26.
Hashizume, H., Ishida, H., and Kohra, K. (1972) *Phys. Stat. Sol. (a)* <u>12</u>, 453.
Tanemura, S. and Lang, A. R. (1973) *Z. Naturforschung* <u>28a</u>, 668.

Multiple crystal monochromators
Beaumont, J. H. and Hart, M. (1974) *J. Phys. E (Sci. Inst.)* <u>7</u>, 832.

Kohra, K. and Kikuta, K. (1968) *Acta Cryst.* A24, 200.
Kov'ev, E. K. and Baturin. V. E. (1975) *Sov. Phys. Cryst.* 20, 8.
Hashizume, H., Iida, A., and Kohra, K. (1973) *Jap. J. Appl. Phys.* 14, 1433.
Matsushita, T., Kikuta, K., and Kohra, K. (1971) *J. Phys. Soc. Japan* 30,
        1136.

Multiple reflections
Okkerse, B. (1963) *Philips Res. Repts.* 18, 413.
Okkerse, B. (1965) *Philips Res. Repts.* 20, 377, 389.

Neutron topography
Ando, M. and Hosoya, S. (1972) *Phys. Rev. Lett.* 29, 281.
Doi, K., Minakawa, H., and Masaki, N. (1971) *J. Appl. Cryst.* 4, 528.
Schlenker, M. and Shull, C. G. (1973) *J. Appl. Phys.* 44, 4181.
Schlenker, M., Baruchel, J., Petroff, J. F. and Yelon, W. D. (1974)
        Appl. Phys. Lett. 25, 382.
Tomimitsu, H. and Doi, K. (1974) *J. Appl. Cryst.* 7, 59.

CHAPTER 3

CONTRAST ON X-RAY TOPOGRAPHS

3.1.   CRYSTALS WITHOUT PLANAR OR LINE DEFECTS

When the X-ray topographic plate shows a uniform intensity across the
crystal, the X-ray topographer is in some degree of difficulty. Is this a
genuinely defect-free crystal or is one not imaging or resolving those defects which are present? In the limits of very high uniform dislocation
density and zero dislocation density we might expect identical images on a
traverse topograph. There are ways of distinguishing between the two cases,
as will be seen subsequently, but fortunately most crystals fall somewhere
between the two extremes. Nevertheless, the interpretation of the features
observed on topographs is extremely difficult in the early stages of one's
career in topography. I hope this chapter will make things a little easier.

Let us consider a situation common to many new workers in topography:
we obtain a topograph showing some features but no recognizable dislocation
lines. We will assume that the guidelines laid down in Chapter 2 have been
followed and to the best of our knowledge the apparatus should be capable of
resolving individual dislocations. What tests should we apply to begin
interpretation? The first has already been mentioned. The rocking curve
is an excellent guide to crystal perfection, as the rocking curve will be
sharp, resolving the $K\alpha_2$ and $K\alpha_1$ peaks with a ratio of 1:2 when the crystal
is nearly perfect and very broad when the crystal is imperfect. A mosaic
spread of a few minutes of arc will be revealed as a broadened rocking curve
in a Lang camera. Secondly, in highly distorted crystals irregular areas
not satisfying a Bragg reflection (white) are often found together with
swirling contrast bands where the distorted reciprocal lattice cuts the Ewald
sphere in various unidentifiable reflections.

There are, however, important clues to the identification of highly
perfect crystals, and one of these is the existence of Pendellösung fringes.
It is rare that a crystal is perfectly flat, and usually there is some tapering at the edges. For example, in silicon, which cleaves on {111} planes,
one often observes thickness fringes at the specimen edge from inclined {111}
surfaces produced by cleavage. The high visibility of Pendellösung fringes
delineating contours of equal thickness, is a very good indication of
crystal perfection because, as we saw in Chapter 1, it is a feature of
"perfect" crystal diffraction. An example of such thickness fringes is
given in Fig. 3.1.

The observation of thickness fringes by Kato and Lang (1959) was one of
the earliest demonstrations that dynamical X-ray diffraction processes were
occurring. Thickness fringes had been predicted by Ewald as early as 1917
and they were known and had been identified in electron microscopy since
1953. However, interpretation in the X-ray case is not trivial. For a
point X-ray source situated 1 metre from a crystal, we find that the angle

63

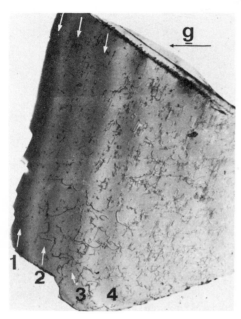

Fig. 3.1. Pendellösung fringes in a
      traverse topograph of a
      wedge shaped crystal of
      silicon. CuKα radiation,
      311 reflection. Field
      width 3.2 mm. Absorption
      damps out the fringes
      after the 4th maximum.

subtended by the first Fresnel zone is
comparable with the perfect crystal
reflecting range. Use of a plane wave
approximation, as in electron microscopy is thus extremely dubious and is
certainly inadequate for shorter
source-specimen distances. In fact,
due to the finite size of the source
and usual operation with substantially
shorter specimen-source distances, the
actual beam divergence is much greater
than the perfect crystal range. In
transmission electron microscopy, the
reflecting range is much larger and
although beam divergence is also
larger, it is seen from Table 3.1 that
it is an order of magnitude smaller
than the reflecting range. Thus in
electron microscopy we find the plane
wave approximation a good one except
in the case of scanning transmission
electron microscopy (STEM) where the
beam divergence is substantially
larger.

   Except in special circumstances,
then, the X-ray beam has a divergence
greater than the width of the dispersion surface diameter and all tie
points along the dispersion surface are
excited. As indicated in Fig. 1.9,
this results in many Bloch waves
travelling in the Borrmann fan. The
intensity in the diffracted beam is
thus dependent on the direction of
energy flow and the phase relations
between the Bloch wavefields.
In particular, in the section topograph, the incident wave can be
represented as a spherical wave and the phase relation across this wavefront
is retained by the wavefields as they propagate in the crystal. However,
we will treat the traverse topograph first when considering Pendellösung
fringes as this, although less fundamental, bears a closer relation to the
electron case.

3.1.1. Pendellösung Fringes in Traverse Topographs

   Takagi (1969) has shown that the intensity on a traverse topograph is
given by the integrated intensity and that this is independent of the
detailed shape of the wavefront. This implies that the phase relations are
destroyed in the traversing, and thus we can use the integrated plane wave
intensity to calculate the intensity of the diffracted beam. This, for the
perfect crystal with zero absorption, can be obtained by integrating (1.52)
and (1.61). In the symmetrical Laue case the integrated diffracted intensity
I_g^T is given by

$$I_g^T = \int_{-\infty}^{\infty} \sin^2 \{\pi t(1 + \eta^2)^{\frac{1}{2}}/\xi_g\}/(1 + \eta^2) \; d\Delta\theta$$

$$= (g\xi_g)^{-1} \int_{-\infty}^{\infty} \sin^2 \{\pi t(1 + \eta^2)^{\frac{1}{2}}/\xi_g\}/(1 + \eta^2) \; d\eta. \qquad (3.1)$$

TABLE 3.1. Validity of the plane wave approximation

Technique	Beam divergence (rad)	Reflecting curve width (rad)
Transmission electron microscopy	10^{-3}	10^{-2}
X-ray topography	5×10^{-4}	10^{-5}

It can be shown that the integral may be written in terms of the zero order Bessel function J_o. Then

$$I_g^T = (\pi/2g\xi_g) \int_0^{2\pi t/\xi_g} J_o(\rho) \; d\rho. \qquad (3.2)$$

This is plotted in Fig. 3.2a. The term outside the integral tells us that high-order long-extinction distance reflections give very low intensity while the integral determines how the intensity varies as a function of thickness. We note that the intensity is an oscillatory function and that the pseudo-period is ξ_g, the extinction distance as defined for $\eta = 0$ in Chapter 1. These oscillations are the Pendellösung fringes observed in a wedge crystal. As the thickness increases, the amplitude of the oscillation decreases as $(\xi_g/t)^{\frac{1}{2}}$ and thus the visibility decreases. When absorption is included, the expression for the intensity becomes cumbersome, but essentially what happens is that the mean intensity decreases with thickness. The general expression has been given by Kato (1961). In the Bragg case for a thick crystal, oscillations are not observed in the integrated intensity. The intensity as a function of thickness is given in Fig. 3.2b.
One of the most exciting features of the discovery of Pendellösung fringes was that here was a method for measuring structure factors without the need for making intensity measurements. Structure factors derived from the fringe spacing across a uniform wedge of known angle are in good agreement with values obtained by classical methods, but the projection topograph is not best suited to such measurements. In current work, section topographs are employed as the fringe visibility is higher and one has more information on the perfection and, importantly, the presence of strains. It is worth noting at this point that in almost all experimental arrangements, both

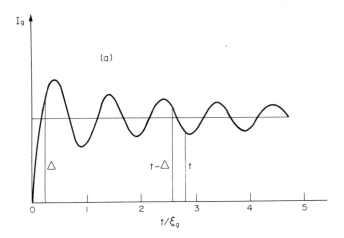

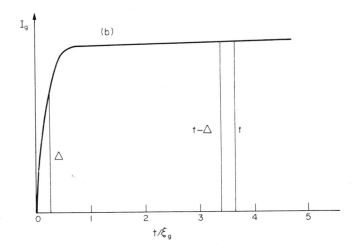

Fig. 3.2. Integrated plane wave intensity as a function of thickness.
        (a) Laue case. (b) Bragg case.

section and traverse, we have two periods, one for each state of polarization. The presence of two periods changes the effective period and also
leads to a fading of the fringes every $\frac{1}{2}(2n + 1)N$ fringes where N is given
by

$$N = \tfrac{1}{2}(1 + C)/(1 - C) \qquad\qquad (3.3)$$

and n is an integer. Polarization effects become very important at large
Bragg angles and should not be overlooked.

3.1.2. Pendellösung Fringes in Section Topographs

Unlike the projection (or traverse) topograph, in section topographs of parallel sided plates, Pendellösung fringes are observed (Fig. 3.3a). The origin can be understood by referring to Fig. 3.3b. The incident wave can be represented as a spherical wave which excites all parts of the dispersion surface coherently. Then we have in any direction AR, two wavefields propagating which correspond to the tie points S and T. These two wavefields interfere. There exist definite phase relations between all the waves in the Borrmann triangle, and clearly as we travel along the base of the Borrmann triangle we will go through points where the two wavefields arriving are successively in and out of phase. Hence we might expect fringes across a section topograph provided that the width of the incident beam is small compared with the width of the Borrmann fan and that the crystal is highly perfect. In crystals which have been irradiated or internally oxidized, small point defect clusters develop which cause appreciable scattering of the wavefields thereby destroying the phase relationships inside the Borrmann triangle. Although too small to be detected in projection topographs, the existence of such defects can be inferred from the loss of visibility in the section topograph Pendellösung fringes (see Fig. 7.2. for an example).

As sketched in Fig. 3.3c, the surfaces of constant phase are on hyperbolic cylinders and thus in a parallel-sided plate the exit surface cuts these cylinders in straight lines; hence the parallel-sided fringes observed in Fig. 3.3a. When the crystal is wedge shaped, the exit surface cuts the hyperbolic cylinders along hyperbolae (Fig. 3.4a) and thus hyperbolic fringes are seen in the topograph. An example of these hook-shaped fringes is given in Fig. 3.4b.

The hook-shaped fringes show a periodic fading and spacing shifts due to polarization effects just as in the projection topographs, and in accurate structure factor measurements these effects must be included. The most serious potential cause of error is that of strain in the crystals. Pendellösung fringes are perturbed by the elastic strains in the crystal, and such contraction of the fringe spacing has serious repercussions on the reliability and reproducibility of structure factor measurements. Much effort has gone into attempts to obtain consistent data for silicon and germanium. For examples of the techniques and difficulties see, for example, Hart and Milne (1969), Kato (1969) and Aldred and Hart (1973).

Clearly in any quantitative interpretation of the contrast in section topographs a spherical wave theory of dynamical diffraction must be employed and such a theory was originally developed by Kato. We will not proceed with the mathematics, and readers are recommended two review articles both by Kato (1963a, 1974).

Before leaving perfect crystal Pendellösung, we note that similar "Kato" fringes have been observed in neutron diffraction, indicating that a spherical wave description of the neutron beam is necessary (Shull, 1973).

3.1.3. Energy Flow in Section Topographs

Section topographs can be very valuable in providing information on the direction of energy flow in the crystal. Defining an energy flow parameter p by

$$p = \tan \theta / \tan \theta_B, \qquad (3.4)$$

we note that this corresponds to a coordinate proportional to the distance
along the base BC of the Borrmann triangle in Fig. 1.9. The extreme values
of p = $\pm$ 1 occur at B and C and p = 0 at the mid point N.

The plane wave intensity in the symmetric Laue geometry, given by
equation 1.62 becomes, via $I^p dp = I^{\eta} d\eta$,

$$I^p = (1 - p^2)^{\frac{1}{2}}. \tag{3.5}$$

This shows that rays in the centre of the Borrmann fan have higher intensity
than those in the margins. However, this does not give us the average
intensity across the section topograph. We must consider the density of
wavefields and we will show that this is least in the centre and greatest in
the margins of the fan.

The angular amplification A is defined as

$$A = d\Theta/d(\Delta\theta), \tag{3.6}$$

where $d(\Delta\theta)$ is the angular divergence of an incident wave-packet and $d\Theta$ is
the angular divergence of the wavefields inside the crystal corresponding to
this elementary wave-packet. Defining the dispersion surface radius of
curvature as R, it is then straightforward to show that

$$A = k \cos \theta_B / R \cos \Theta. \tag{3.7}$$

We see that very close to the Bragg condition, where the dispersion surface
is highly curved, R << k and the crystal acts as a powerful angular
amplifier. For silicon 220 reflection with MoKα radiation, A reaches
3.5×10^4 in the centre of the dispersion surface. Far from the centre, the
dispersion surface becomes asymptotic to the spheres about the reciprocal
lattice points and A decreases to unity. Thus when the whole of the dispersion surface is excited by a spherical wave, due to the amplification
close to the Bragg condition, the density of wavefields will be very low in
the centre of the Borrmann fan and extremely high in the margins.
Calculations of the intensity, averaged over the spherical wave Pendellösung
oscillations, compounding the intensity and density of wavefields, have been
performed by Kato (1960). His calculations include absorption, and his
results are reproduced in Fig. 3.5.

The very important feature of the low absorption curve is that the diffracted beam intensity rises very sharply at the extremal values p = $\pm$ 1,
that is at the edges of the section pattern. As we have seen, it arises
from the very high density of wavefields in these regions. The increase in
intensity at the edges of Laue spots from some crystals had been known to
classical crystallographers for many years but it had been atributed to surface damage. The observation of the 'hot margin' in the X-ray section topographs conclusively demonstrated it to be a dynamical diffraction effect.
Under low absorption conditions, the presence of a hot margin is a good
indication that the crystal is highly perfect. Point defect clusters, for
example from radiation damage, will destroy the phase relationships and destroy the margin effect. However, Lang (1970) takes some pains to point out
that, under moderate absorption conditions, $\mu t \simeq 2$, the perfect crystal
profile is very similar to that of a crystal containing a large amount of
damage. In interpreting such topographs, moderate absorption conditions
should be avoided, advice which is of fairly general validity in topography.

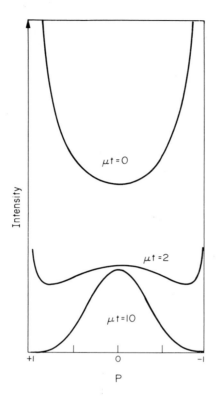

Fig. 3.5. Calculated intensity profile across the diffracted beam
         of the section topograph for various absorption
         conditions (after N. Kato).

3.2.   DYNAMICAL DIFFRACTION IN DISTORTED CRYSTALS

3.2.1. Small Distortions

When the deformation is small, theories analagous to the "Eikonal" theory
in optics are applicable. The concept of a ray is retained and the crystal
deformation is described by a continuously varying reciprocal lattice
vector. In optics the corresponding situation occurs when light propagates
through a medium of slowly varying refractive index. The original modes of
propagation in the undeformed crystal retain their identity, in the X-ray
case as Bloch wavefields, but in order to accommodate the deformation each
mode undergoes a gradual transformation. We will proceed to develop the
Eikonal theory of Penning and Polder (1961) following closely the treatment
given by Malgrange at the 1975 Limoges Summer School.
In a distorted crystal, where the atomic displacement from the perfect
crystal is given by the vector $\underline{u}$, we can define a local reciprocal lattice g'

by

$$\underline{g}´ = \underline{g} - \text{grad}(\underline{g}.\underline{u}) \qquad (3.8)$$

Let us consider two points F and G between which the local reciprocal lattice varies from g´ to g´ + dg´. When the deformation is small the shape of the dispersion surface does not change and only a displacement of the hyperbolae results. We can consider this as a rotation about the origin of reciprocal space, and then the dispersion surface moves along the asymptote (Fig. 3.6) by a vector given by $\underline{\nu}$, where

$$\underline{\nu}.\underline{K}_o = 0, \qquad \underline{\nu}.\underline{K}_g = \underline{K}_g.d\underline{g}´. \qquad (3.9)$$

When $\underline{K}_g . d\underline{g}´ = 0$ no movement of the dispersion surface takes place and thus for a local reciprocal lattice where the vectors dg´ are always perpendicular to $\underline{K}_g$, the wavefields remain unchanged. In the general case, the loci of points where dg´ is perpendicular to $\underline{K}_g$, form surfaces along which the dispersion surface remains static and the associated wavefields unchanged. These surfaces divide the crystal up into domains which can be considered as perfect and are analagous to the surfaces of constant refractive index in the optical case. As each domain will be misoriented or dilated with respect to its neighbour, the wavevector $\underline{K}_o$ changes by $d\underline{K}_o$ on crossing the domain boundary. In exact analogy with the optical case this is given by

$$d\underline{K}_o = \gamma\text{grad}(\underline{g}´.\underline{K}_g) = -\gamma\text{grad}(\underline{K}_g.\text{grad})(\underline{g}.\underline{u}) \qquad (3.10)$$

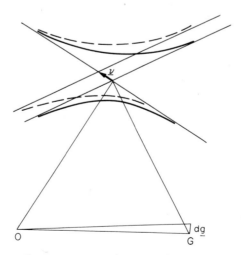

Fig. 3.6. Dispersion surface construction showing how local distortion leads to a shift of the dispersion surface along one of its asymptotes.

The constant γ may be found in the following manner. The fundamental equations of the dispersion surface given in (1.11) are

$$K_o^2 = k^2(1 + \chi_o) + k^2 C \chi_g R,$$

$$K_g^2 = k^2(1 + \chi_o) + k^2 C \chi_{\bar{g}}/R. \tag{3.11}$$

From this we derive, for $\chi_g = \chi_{\bar{g}}$,

$$d\underline{K}_o \cdot (\underline{K}_o + R^2 \underline{K}_g) = -R^2 \underline{K}_g \cdot d\underline{g}'. \tag{3.12}$$

If $d\underline{r}$ is the path vector between F and G

$$d\underline{g}' = (d\underline{r} \cdot \mathrm{grad})\underline{g}'. \tag{3.13}$$

Defining a vector $\underline{P}'$ parallel to the Poynting vector given by

$$\underline{P}' = \underline{K}_o + R^2 \underline{K}_g \tag{3.14}$$

we find

$$\gamma = - R^2 d\ell/|\underline{P}'| \tag{3.15}$$

where $d\ell$ is the path length along the ray trajectory and given by

$$d\underline{r} = \underline{P}' d\ell/|\underline{P}'| \tag{3.16}$$

We can now use (3.11) to find the variation of R with respect to $\underline{g}'$. This is

$$dR = (2R^2 d\ell/k^2 C \chi_g |\underline{P}'|)(\underline{K}_o \cdot \mathrm{grad})(\underline{K}_g \cdot \mathrm{grad})(\underline{g} \cdot \underline{u}) \tag{3.17}$$

which may be written as

$$dR = \frac{2R^2 d\ell}{C\chi_g |\underline{P}'|} \frac{\partial^2 (\underline{g} \cdot \underline{u})}{\partial s_o \partial s_g} \tag{3.18}$$

We have now obtained an important result. Instead of considering the dispersion surface as a variable and the reciprocal lattice as invariant, it is usually easier to consider the reciprocal lattice as the variable. Then (3.18) determines the variation of the amplitude ratio of the reflected and transmitted components of a wavefield propagating through the crystal.

Let us see what this implies. The ratio R characterizes a particular tie point on the dispersion surface. Thus if R varies the tie point must migrate along the dispersion surface branch. As the direction of energy flow, i.e. the ray direction, is everywhere normal to the dispersion surface, the ray direction varies as the tie point migrates. The rays

therefore propagate along curved paths. We further note that as the value
of R varies, so does the intensity in the diffracted and transmitted beams.
The important parameter in determining the contrast is thus the integral
along the ray path of the term $\{\partial^2/\partial s_o \partial s_g\}(\underline{g}\cdot\underline{u})$. The ray path parameter
β is defined as

$$\beta = (kC\chi_g)^{-1}\{\partial^2/\partial s_o \partial s_g\}(\underline{g}\cdot\underline{u}). \tag{3.19}$$

From (3.18) it can be seen that the sign of the change in R is independent
of the sign of R and depends only on β. As can be clearly seen in Fig. 3.7,

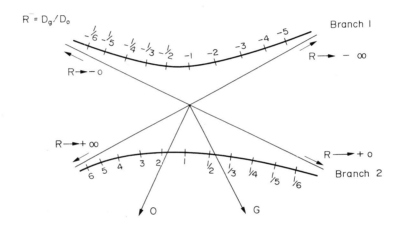

Fig. 3.7. Amplitude ratios on the two branches of the dispersion
          surface.

the sign of R is opposite for the two branches of the dispersion surface.
However, on both branches R increases with η and thus for a given
deformation, the tie points on both branches of the dispersion surface
migrate in the same direction. The curvature of the ray corresponding to
branch 1 is in the same sense as the curvature of the reflecting planes
while the curvature of branch 2 rays is of opposite sense to the lattice
curvature. This has been experimentally verified by Hart and Milne (1971).
    Before looking at some applications of the Eikonal theory it is
important to investigate the range of validity of the theory. As stated
previously, the distortions must be small if the concept of a ray is to be
retained. Penning (1966) has shown that the theory is applicable provided
that the radius of curvature of the reflecting planes does not exceed a
critical value R_c. This is approximately equivalent to an angular rotation
of the Bragg planes by half the reflecting curve width in an extinction
distance. The critical radius of curvature is thus

$$R_c \simeq g\xi_g^2 \qquad\qquad (3.20)$$

Kato (1963b) has shown that on such a criterion the ray theory breaks down approximately 10 μm from a dislocation core. Unfortunately, then, we cannot apply the Eikonal theory to compute the contrast of a dislocation on an X-ray topograph in a quantitative way. However, as we shall see, it is possible to obtain a large amount of qualitative information from such an approach.

The Eikonal theory of X-ray diffraction was originally developed by Penning and Polder (1961) for the case of anomalous transmission in a distorted crystal. Here we have a relatively simple situation because, as a result of the Borrmann effect, only the branch 1 wavefields close to the exact Bragg condition are present. The energy flow here is parallel to the lattice planes. Let us consider first a dislocation in the centre of a thick crystal and look at the contrast a long way from the dislocation core (Fig. 3.8). On one side of the defect, the curvature is first positive and

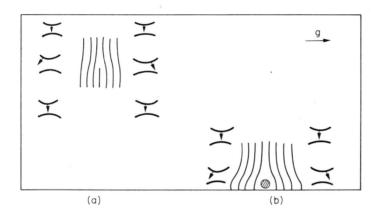

(a)   (b)

Fig. 3.8. Migration of tie points in the distorted regions around
   (a) a dislocation in the centre of the thick crystal.
   (b) a precipitate close to the exit surface.

then negative while the converse is true on the opposite side of the defect. Thus the tie point migration is in opposite sense on either side of the dislocation line, but, provided the distortion is small so that the tie points retain their identity, the migration above and below the defect exactly compensates on both sides of the dislocation. At first sight we would expect there to be no net contrast, as, in fact, would be the case if no absorption was present. However, contrast is observed in the following way. Close to the exact Bragg condition (which corresponds to our branch 1 anomalously transmitted wavefield) the angular amplification is extremely large (3.7), and hence very small distortions of the Bragg planes lead to appreciable changes in the direction of energy flow. The rays so deviated suffer additional absorption because the effective absorption coefficient is sensitive to the deviation from the exact Bragg condition. Thus there is a loss of intensity in the rays coming from both sides of the dislocation, and the long range image of the dislocation appears white.

When the defect is close to the surface of a thick crystal, however, we may observe opposite contrast on either side of the defect. Take, for example, a precipitate compressing the surrounding lattice as sketched in Fig. 3.8. Consider first the right hand side of the defect. The ray path parameter β is first negative and the tie point of the branch 1 wavefield migrates to the right, following the lattice plane curvature. At the surface the lattice planes emerge normally, and when the defect is close to the surface the lattice planes must rotate very rapidly. Thus the curvature is very large — too large for the Eikonal theory to be applicable and the rays pass out of the crystal without further diffraction. The contrast is mainly given by the amount of migration taking place in the negative curvature region above the defect where the lattice planes are only mildly distorted. Qualitatively, we can see from Figs. 3.7 and 3.8 that the migration causes an increase in the intensity in the diffracted beam. Conversely, on the left-hand side of the defect we find decreased intensity. A precipitate close to the surface shows black-white contrast which is enhanced on the side of positive g when the surrounding lattice is under compression. An example is shown in Fig. 3.9.

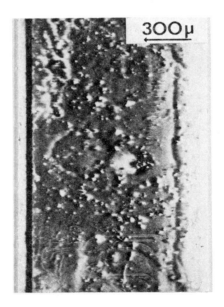

Fig. 3.9. Precipitates in a rare earth vanadate crystal close to the exit surface showing asymmetric contrast under high absorption conditions. (μt = 25). The surrounding lattice is under compression. Scale arrow indicates the direction of the direction of the diffraction vector.

We note here that the black-white contrast reverses with the diffraction vector and also with the sense of the strain in the lattice. This is a useful means of determining the nature of a precipitate. On the side of positive g, if the contrast is enhanced, the lattice is under compression, if reduced, it is under tension. This rule was formulated empirically by Meieran and Blech (1965).

These ideas may be quantified somewhat, and Lawn (1968) has used such an analysis to measure the force per unit length, analogous to the surface tension, in a strip of aluminium evaporated onto a silicon single crystal substrate. Using a simple model for the elastic distortion below the strip, he calculated the maximum radius of curvature of the Bragg planes below the strip edge. Only when the curvature exceeds R_c given in (3.20) is contrast observed, as otherwise the tie point migration effects exactly compensate. By fitting the measured width of the contrast band on the topographs with the elastic model the force per unit length in the surface can be obtained.

It is important to note that the discussion above relates to crystals which are sufficiently thick to consider only one branch of the dispersion surface active. In thin crystals, contrast is observed, via a somewhat different mechanism, from those X-rays not satisfying the Bragg condition, and, as we will see later, the contrast in the so-called "direct image" is symmetric. The useful asymmetry of the image occurs only in absorbing crystals.

Penning-Polder theory has been used by Hart (1963) to compute the long-range contrast from dislocations. As we have noted, the contrast is sensitive to the lattice curvature and provides a means of determining the sense of the Burgers vector. Dislocation contrast has also been studied by Kambe (1963) and Tanner (1973) using the Penning-Polder theory.

A large amount of information can be extracted from topographs by applying the Penning-Polder theory in a qualitative manner such as illustrated above. However, if quantitative information is required about small distortions, then the amount of tie point migration is insufficient. We must also compute the ray trajectory. The information is, in fact, contained in (3.18). By substituting the energy flow parameter p defined in (3.4) for R, one can show that

$$\pm \frac{d}{dz} \left\{ \frac{p}{(1 - p^2)^{\frac{1}{2}}} \right\} = \frac{1}{kC\chi_g \cos \theta_B} \frac{\partial^2 (g \cdot u)}{\partial s_o \partial s_g} \tag{3.21}$$

As has been pointed out by Kato, this is identical to the equation of motion in relative mechanics. In principle, then, we can use (3.21) to compute the ray trajectory and then determine the amount of tie-point migration by (3.18). We thus know both the amplitude and the position of the rays emerging from the exit surface of the crystals, and hence the intensity at any point in the topograph is known. Of course, except in the simplest circumstances, use of a high-speed digital computer will be necessary to perform the calculations.

Kato (1963b, 1964) has developed a very similar theory extended to the spherical wave case. Instead of calculating the tie point migration in terms of the amplitude ratio R, the phase change along the ray for the two components of the wavefield is computed. There is, however, no difference in the results as all "ray" theories rely on the same geometrical optical concepts. In some important recent papers, Kato, Patel and Ando have computed the contrast in a section topograph of a crystal with a superposed oxide film. At the film edge the distortion introduced additional phase changes in the rays, and thus the section topograph fringe pattern (Fig. 3.4) is disturbed. Excellent agreement has been found between computed and experimental topographs. The force per unit length of line as a function of film thickness was obtained by fitting experimental data with theory and it was shown that the stress in the film was constant even for oxide films grown under different conditions and at different temperatures (Kato and Patel, 1973; Patel and Kato, 1973; Ando, Patel, and Kato, 1973).

3.2.2. Large Distortions

   In regions of high distortion the Eikonal theory becomes invalid as it is no longer possible to define a ray. We can consider that interbranch scattering or tie-point jumping occurs when a wavefield enters a highly distorted region. This is illustrated in Fig. 3.10. When the Bloch wave enters

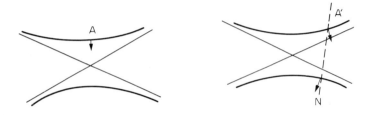

Fig. 3.10. Tie point jumping (or interbranch scattering) of wave-
          fields on passing through a heavily distorted region of
          crystal. The original tie point A migrates to A' and then
          excites a new wavefield corresponding to N.

distorted material, the tie-point A migrates until a critical radius of curvature R_c is reached and it then excites a new tie-point N on the opposite branch of the dispersion surface. This is equivalent to saying that the wavefields decouple on crossing the heavily distorted region and excite new wavefields in the less-deformed region below. This type of approach works well for calculating the contrast of a planar defect (Authier, 1968) and is very useful in obtaining a qualitative understanding of dislocation contrast. However, it is not useful for quantitative synthesis of dislocation images.
   To do this we need to go to the generalized form of the dynamical theory introduced in section 1.8. The approach is very similar to that employed by electron microscopists in computing dislocation images with two important complications. Firstly, X-ray topographers cannot approximate the incident wave to a plane wave. Secondly, in transmission electron microscopy the Bragg angles are very small and the base of the Borrmann fan is very narrow. One can then make a very powerful approximation known as the "column approximation" in which the crystal is divided into columns parallel to the beam direction which are considered independently. In the X-ray case this is not permissible, and although Howie and Basinski (1968) have developed a theory which takes into account the scattering between neighbouring columns

for use in weak beam electron microscopy, it is preferable not to make the
approximation at all in the X-ray case.

Takagi's equations, or the equations developed independently by Taupin
(1964), are capable of quantitative application to defect contrast in topography. In the distorted crystal the wavevector $\underline{K}'_g$ differs from the ideal
crystal wavevector by $\Delta\underline{g}$ where

$$\Delta\underline{g} = -\text{grad}(\underline{g}.\underline{u}) \qquad\qquad (3.22)$$

and $\underline{u}$ is the atomic displacement. The β'_g parameter then is a function of
position in the crystal. The electric displacement in the crystal has a form

$$\underline{D}(\underline{r}) = \sum_g \underline{D}'_g(\underline{r}) \exp(-2\pi i(\underline{K}_g.\underline{r} - \underline{g}.\underline{u})). \qquad\qquad (3.23)$$

Takagi's equations are then

$$\partial D'_o/\partial s_o = -i\pi kC\chi_{\bar{g}}D'_g,$$

$$\partial D'_g/\partial s_g = -i\pi kC\chi_g D'_o + i2\pi k\beta'_g D'_g. \qquad\qquad (3.24)$$

These may be combined to give two equations of the form

$$\partial^2 D'_o/\partial s_o\partial s_g - i2\pi k\beta'_g\partial D'_o/\partial s_o + \pi^2 k^2 C^2\chi_g\chi_{\bar{g}}D'_o = 0,$$

$$\partial^2 D'_g/\partial s_o\partial s_g - i2\pi k\beta'_g\partial D'_g/\partial s_o + (\pi^2 k^2 C^2\chi_g\chi_{\bar{g}} - i2\pi k\partial\beta'_g/\partial s_o)D'_g = 0.$$

$$(3.25)$$

These have hyperbolic form, and in a perfect crystal can be solved by the
method of Riemann. In a distorted crystal Takagi's equations are capable of
numerical integration. The amplitudes of the waves through the crystal may
be evaluated using a grid such as illustrated in Fig. 3.11.

If M is the point at which we require the wave amplitude and P and Q are
neighbouring points at which the amplitudes are known, Authier, Malgrange,
and Tournarie (1968) have shown that the equations can be written

$$D_o(M) = D_o(P) + p(-i\pi kC\chi_{\bar{g}})D_g(P),$$

$$D_g(M) = D_g(Q) + q(-i\pi kC\chi_g)D_o(Q) + q(2\pi ik\beta_g)D_g(Q), \qquad (3.26)$$

where p = PM and q = QM as defined in Fig. 3.11. It is therefore straightforward, if a little time-consuming, to derive the amplitudes at any point
in the crystal by a step-by-step calculation.

Several workers (e.g. Balibar and Authier, 1967) have now used such
computer integration methods to generate dislocation images in section
topographs. As illustrated in Fig. 3.12, the agreement between the experimental and simulated images is excellent. Marked differences can be seen in
the contrast of the section pattern when the magnitude or the sense of the
Burgers vector is changed (Epelboin, 1974), and this provides a powerful tool
for determining both these quantities. Routine generation of images using

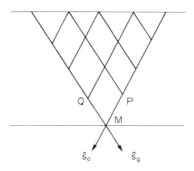

Fig. 3.11. Grid used for the numerical evaluation of Takagi's
           equations in a distorted crystal.

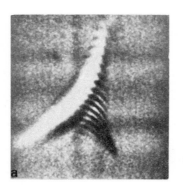
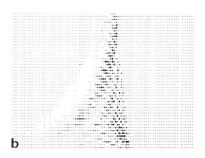

Fig. 3.12. (a) Experimental image of a dislocation in a section
           topograph (courtesy F. Balibar). (b) Computer simula-
           tion of the image (courtesy Y. Epelboin).

overprinting techniques is now possible (Epelboin and Lifchitz, 1974), even
if it is necessary to use a somewhat arbitrary grey scale.
    The wave theory is clearly very successful at simulating dislocation con-
trast on section topographs, but it is aesthetically rather unsatisfactory.
The concept of the ray is totally lost inside the crystal and several workers
have studied the relationship between the Eikonal and wave theories (e.g.
Kato, 1973). Recently, Balibar and Malgrange (1975) have attempted a new
approach based on the concept of wave packets and there are indications that
a theory of this kind may give a better physical insight into the processes
taking place inside the crystal.
    Finally , we note that in the limit of small Bragg angles, where $\hat{s}_o$ and $\hat{s}_g$

can be considered to be parallel, Takagi's equations reduce to the Howie-
Whelan equations, well known to electron microscopists.

3.3.   CONTRAST OF CRYSTAL DEFECTS IN TOPOGRAPHS

Two important review articles have been published by Authier (1967, 1970)
and these will repay detailed study.

3.3.1. Dislocations in Section Topographs

Dislocation images in X-ray topographs differ from those in electron
micrographs primarily because the incident wave must be regarded as a
spherical wave rather than a plane wave. As shown in Fig. 3.13, we suppose
that the dislocation line D cuts the extrema of the Borrmann fan at S and T.

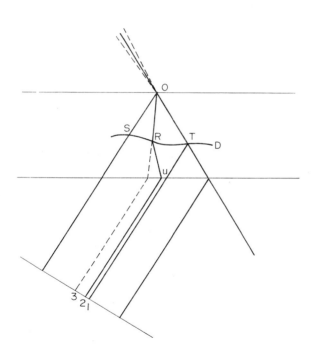

Fig. 3.13. Formation of the three types of
          image in section and projection
          topographs: 1, Direct image,
          2, Intermediary image,
          3, Dynamical image (after
          A. Authier).

Three types of image have been
characterized by Authier
(1967, 1970). At T the dislocation cuts the beam of
X-rays which are not diffracted by the crystal. As the
X-ray source is polychromatic,
most of the intensity will fall
into this category. In the
highly deformed region around
the dislocation, X-rays with
angular deviation from the
exact Bragg angle of less than
the perfect crystal reflecting
range will not be diffracted,
but some rays from the polychromatic "direct" beam will.
Further, these X-rays will be
diffracted only in the deformed region around the dislocation and not in the perfect
crystal. They thus suffer no
primary extinction and the so
called "direct image" appears
as an intense dark spot on the
section topograph. In thick
crystals the rays far from the
Bragg angle contributing to
the direct image suffer ordin-
ary absorption and hence the
direct image does not appear.
Within the Borrmann fan two
types of image arise.
Consider two wavefields propagating along OR. On crossing
the dislocation at R the two
waves decouple into their
transmitted and diffracted

components. When they re-enter perfect material they excite new wavefields.
Intensity is removed from the direction OR and the dislocation casts a
shadow known as the "dynamical image". This is always less intense than
background. The newly created wavefields propagate along paths such as RU,
and these give rise to the third type of image - the "intermediary image".
As the newly created wavefields interfere with the original wavefields, the
intermediary image shows an oscillatory contrast. These are clearly visible
inside the "banana"-shaped dynamical image in Fig. 3.12. The Borrmann fan is
filled with wavefields propagating in different directions, and thus the
dynamical and intermediary images have much poorer spatial resolution than
the direct image.

3.3.2. Dislocations in Traverse Topographs

   In principle we can understand the traverse topograph contrast by considering a series of overlapping section topographs. While this can be useful
it is sometimes difficult to relate the two, though we see quite generally
that the dynamical and intermediary images will blur out and information
will be lost.
   Takagi (1969) has shown that the traverse pattern corresponds to that
produced by an integrated plane wave and we can approach dislocation images
in traverse topographs from this angle. In low absorption conditons where
$\mu t < 1$, dislocations in Lang topographs appear as regions of enhanced
intensity (black on the plate). These are direct images formed by X-rays not
satisfying the exact Bragg condition in the perfect crystal. The slope of
the intensity versus thickness curve in the Laue (transmission) case is
everywhere less than its value at the origin (Fig. 3.2a). Thus the intensity
I_t from a perfect crystal thickness t is always less than that from perfect
material thickness $t - \Delta$ and a deformed region thickness Δ (Fig. 3.14).
That is

$$I_t < I_{t-\Delta} + I_\Delta \tag{3.27}$$

as the X-rays diffracted from the distorted region diffract kinematically and
suffer no extinction. The best direct image contrast is observed at the
first Pendellösung minimum at a thickness 0.88 ξ_g (Lang, 1970).

   In the Bragg (reflection) geometry the integrated intensity is independent
of crystal thickness for crystals thicker than about an extinction distance
(Fig. 3.2b). Thus the intensity of the defect plus the intensity from a
perfect crystal of thickness $t - \Delta$ is always greater than that from the
perfect crystal of thickness t. Direct images in reflection topographs are
also more intense than the background.
   The direct image is considered to be formed when the effective misorientation of the lattice planes, given by

$$\delta(\Delta\theta) = -(k \sin 2\theta_B)^{-1} \, \partial(\underline{g}.\underline{u})/\partial s_g \tag{3.28}$$

is greater than α times the perfect crystal reflecting range, where α is
approximately unity (Authier, 1967). Generalized cylinders of misorientation
may be drawn around a dislocation line using continuum elasticity theory and
the full width calculated from the projected width of the cylinder circum-
scribed by the contour where $\delta(\Delta\theta)$ is equal to α times the reflecting range.
As the intensity in the direct image is proportional to the volume of

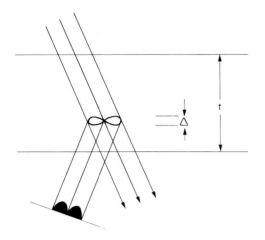

Fig. 3.14. Cylinders of misorientation
around an edge dislocation.
The thickness of material
having misorientation
greater than the perfect crystal range is denoted Δ. The
two parts of the image arise
from the two lobes of opposite
misorientation.

"mosaic" crystal contained within the cylinder, the direct image may be synthesized. It is important to note that any inter-branch scattering and interference effects taking place as a result of the insertion of the mosaic region into the perfect crystal are neglected. The mosaic diffracting region model predicts the direct image only and, for example, gives no explanation for the asymmetric profiles of dislocations often found in Lang topographs. Miltat and Bowen (1975) have recently performed a detailed study of many direct images and concluded that the mosaic model is more nearly applicable than the 'diffracting zone' model based on the computation of misorientation gradient contours. The value of α was also found to be a function of the reflection used.

A very useful working expression for the width of a direct image based on the mosaic model has been given by Lang and Polcarova (1965). The width W of a dislocation image is given approximately by

$$W \simeq \xi_g (\underline{g} \cdot \underline{b})/2\pi .$$
(3.29)

Use of this expression enables a quick check to be made on Burgers vector asignments. The mosaic model explains the significance of the bimodal pro-file of dislocations when $\underline{g} \cdot \underline{b} > 2$. Lobes of opposite misorientation give rise to two components of the image (Fig. 3.14). The distance between the two maxima is inversely proportional to the reflecting curve width, and for low-order reflections direct images usually appear single due to overlap of the two components. In high order reflections, double images are common and the peak separations are found to be in good agreement with the predictions of the model.

On the ideas developed here, direct images are expected to be invisible in very thin crystals where the kinematical approximation is satisfied (Penning and Goemans, 1968). The intensity is then proportional to the thick-ness and

$$I_t = I_{t-\Delta} + I_\Delta$$
(3.30)

The exact thickness at which dislocations disappear is a function of reflection vector (Tanner, 1972) but is about 0.4 ξ_g for low-order reflections, falling for higher order reflections.

Dynamical images, although found in thin crystals, are generally important in thick crystals. As in section topographs, they are less intense than
the background and appear white on the plates. In thick crystals where only
the branch 1 wave remains with $\eta = 0$, the presence of the dislocation causes
scattering of intensity out of that wavefield into newly created wavefields.
All these suffer considerable absorption except those on branch 1 close to
$\eta = 0$, and thus the dislocation casts a shadow in both diffracted and transmitted beams. In some cases interference can be observed between the newly
created wavefields leading to oscillatory contrast in the region of the dislocation close to the exit surface (the intermediary image). Only when the
extinction distance is short are such effects usually observed as anomalous
absorption quickly damps out the branch 2 waves within a short distance from
the exit surface. Where they intersect the exit surface, dislocations often
exhibit a black-white contrast from the surface relaxation which can be
interpreted using the Penning-Polder theory along the lines discussed in
section 3.2.1. Due to divergence of the wavefields in the crystals, dynamical images have best contrast when the dislocation is close to the exit
surface.

Intermediary images are usually blurred out by the traversing but often
significantly effect the shape of the direct image leading to asymmetry of
the profile. In suitable circumstances the intermediary image is well
separated and, as in the section topograph, shows oscillatory contrast. The
splendid intermediary images of Authier (1967) are rare, and more often the
overlap of direct and intermediary images leads to oscillatory contrast of an
inclined dislocation (Miltat and Bowen, 1975). Examples of direct, dynamical,
and intermediary images are given in Fig. 3.15 abc.

3.3.3. Contrast of Precipitates

Provided the long range strain is sufficient to cause a change in $\delta(\Delta\theta)$
greater than the perfect crystal reflecting range, precipitates can be
observed on topographs. A spherical precipitate has a radial strain field
and this forms a characteristic image. It consists of two lobes separated by
a line of no contrast which is always perpendicular to the diffraction
vector. In this direction the atomic displacements are parallel to the Bragg
planes which, hence, suffer no distortion in the g direction. Often
precipitate contrast is more complex as the strain field is often non-radial.
Under high absorption conditions, precipitates often show black-white contrast, which can be explained by the Penning-Polder theory. A striking
example of contrast from a radial strain centre is given in Fig. 3.15 (d).
Normally precipitates show only two lobes separated by a line of no contrast.
This one shows four and under low moderate absorption the image is asymmet-
rical and we see that the sense of the distortion changes sign. Far from
the centre the stress is tensile while close to it the stress is compressive.
One possible explanation is that during device fabrication a pinhole in the
oxide film allowed boron to diffuse into the silicon, resulting in a tensile
stress. During subsequent treatment, aggregation of impurity resulted in a
precipitate at the centre giving a shorter range compressive stress. This is,
however, speculative.

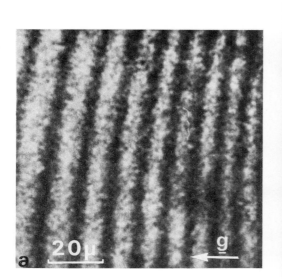

(Fig. 3.15)

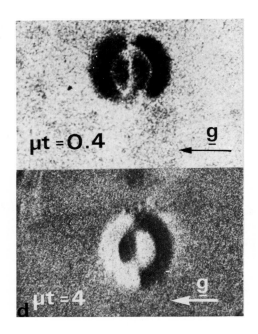

3.3.4. Surface Damage

This is a feature of topography which the novice meets
early. Due to the extreme strain
sensitivity of topography, a
strain-free surface must be
prepared or the contrast will
obscure details from the interior.
Note that the emphasis is on a
strain-free surface - not a flat
one. Scratches on the surface
are only imaged when appreciable
strain is associated with them.

In silicon, for example, there
is a tendency to propagate micro-
cracks below the surface when
mechanically polishing, and an
optically perfect surface will
show a network of scratches on
topographic examination. Some
form of chemical or electrochemical polishing is essential.
Silicon chips for device fabrication are frequently only lapped
and chemically polished on one
surface only, and an etch in
$HF-HNO_3$ to produce the characteristic "orange peel" effect is
necessary prior to topography.
"Precipitates" seen in an
unpolished sample may well be
damage from cutting with a diamond wheel. "Syton" (colloidal
silicon) polishing of hard
materials such as silicon or GGG
seems to be satisfactory without
subsequent chemical polishing as
"Syton" contains NaOH which acts
to relieve the strain during the
mechanical abrasion.

Fig. 3.15. (a) Direct images of dis-
     locations in silicon. The
     narrow images result from
     overlapping of adjacent strain-
     fields in the array. CuKα rad-
     iation, 111 reflection.
     (b) Dynamical images D in a
     thick crystal of $TmVO_4$.

Intermediary images I occur
close to the outcrop of the
dislocations with the exit
surface. 200 reflection, MoKα
radiation, field width 1.3 mm.
(c) Dynamical images of dislocations running parallel to
the crystal surface (courtesy
A.D. Milne). (d) Spherical
strain centre showing anomalous double lobe contrast.
Silicon, 220 reflections.

3.3.5. Contrast of Stacking
        Faults in Section Topographs

Stacking fault contrast in
section topographs can be most
easily understood by expanding the
incident spherical wave as a
Fourier expansion of plane waves
using the method of Kato. It is
also possible to use Takagi's theory to calculate the contrast, but the
physical insight is lost.

Following the first approach, we consider a stacking fault inclined to the crystal faces. Figure 3.16a represents a section, cut in the plane of

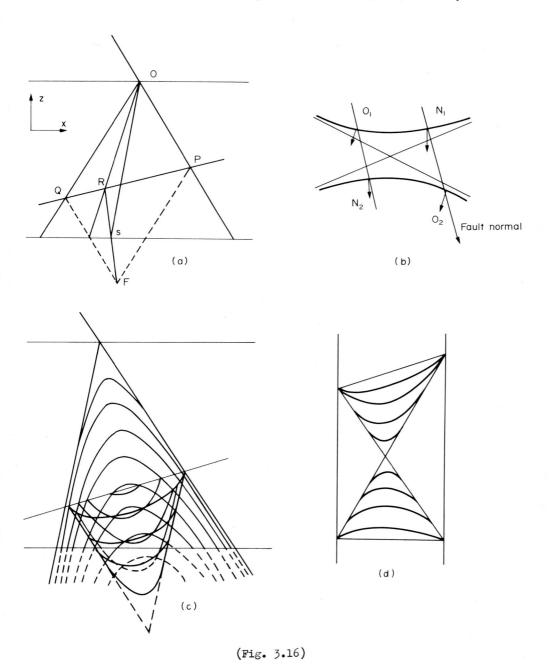

(a)

(b)

(c)

(d)

(Fig. 3.16)

Fig. 3.16. (a) Ray paths for original and new wavefields in a
          crystal containing a stacking fault. (b) Tie points
          O_1 and O_2 corresponding to original wavefields propa-
          gating along OR. Tie points N_1 and N_2 correspond to
          the new wavefields propagating along RS. (c) Surfaces
          of constant phase. The bolder-drawn contours are the
          phase fronts corresponding to the interference between
          the original and the new wavefields. (d) Flat fringes
          in the "hour-glass" formed by an inclined stacking
          fault. The flat fringes correspond to the interference
          phase fronts illustrated in (c).

incidence, through the Borrmann fan. The fault cuts the extrema of the fan
at P and Q. Above the fault two wavefields corresponding to the tie points
O_1 and O_2 in Fig. 3.16b propagate in the direction OR. On crossing the
fault they decouple into their plane wave components, but on re-entering the
perfect crystal below the fault they immediately excite new wavefields. The
boundary conditions can be applied to the plane wave components in a similar
manner to those at a vacuum interface provided that the phase difference
introduced by the fault is included. We see from Fig. 3.16b that four waves
now exist in the crystal. Two of these correspond to the original tie points
O_1 and O_2 and two to newly created Bloch wavefields corresponding to the
newly excited tie points N_1 and N_2. Thus at a point S on the exit surface
the intensity is made up of contributions from the newly created wavefields
travelling along RS and the original wavefields travelling along OS.
Explicitly, the intensity comprises three components, I_1, I_2 and I_3 given by

$$I = I_1 + I_2 + I_3 = |D'|^2 + |D''|^2 + 2R(D'^*D'') \qquad (3.31)$$

where D' is the amplitude of the original wavefields and D'' represents the
amplitude of the new wavefields. I_1 then represents the interference
between the original wavefields which gives rise to the section topograph
Pendellösung (or Kato) fringes, I_2 corresponds to the interference between
the new wavefields and I_3 gives the contribution due to interference between
original and new wavefields. Expressions for I_1, I_2 and I_3 in the general
case of the absorbing crystal have been given by Authier (1968). In the
case of a highly absorbing crystal, I_3 dominates as the contrast arises from
the interference of waves associated with tie points O_1 and N_1. These are
both on branch 1 and are thus only weakly absorbed.
    In the region OPR interference of the original wavefields occurs and the
fronts of equal phase, shown in Fig. 3.16c, are hyperbolic. In the triangle
QPF the newly created wavefields propagate and the phase fronts are also
hyperbolae, equivalent to the phase fronts in OPQ reflected about PQ.
In addition, the flat, non-hyperbolic phase fronts due to the interference
of the two systems occur in QPF. These are shown as the bolder-drawn contour
in Fig. 3.16c. The newly created wavefields converge to a focus F and we
note that QF = OP and PF = OQ. As the line of intersection PQ moves through

the crystal, so the position of the focus F varies. In the particular case where PQ is parallel to the crystal surface, it is easy to see that when PQ lies exactly halfway through the crystal, F is at the exit surface. Quite generally there is a position where no fringes from I_2 and I_3 are observed and the resulting section topograph from an inclined fault has the delight-ful "hour-glass" shape illustrated in Fig. 3.16d. The flat fringes due to interference between original and new wavefields have twice the periodicity of the hyperbolic fringes.

In non-absorbing crystals, the hour-glass contains two sets of fringes, one of each periodicity, but in thicker crystals the I_2 term becomes vanish-ingly small. The contrast of the I_3 fringes is always higher than that of the I_2 fringes, and it is these non-hyperbolic fringes which are usually observed experimentally. A total of t/ξ_g flat fringes is found.

Under the classification of Amelinckx, stacking faults give rise to α fringes. The structure in the region below the fault is derived from that above by a simple translation $\underline{R}$ parallel to the fault (Fig. 3.17a). In con-trast, across a twin boundary the region below is related by a displacement vector $\underline{R}$ which increases linearly from the boundary (Fig. 3.17b), and the fringes thus produced are known as δ fringes. Reverting to the stacking fault we anticipate that the three intensity terms might depend on the value of $\alpha = 2\pi\underline{g}.\underline{R}$, that is, the phase difference introduced into the wavefields by the stacking fault. The terms I_1 and I_2 depend only on $\sin^2 \alpha/2$ (Authier, 1968) and are insensitive to the sign of α. In zero absorption conditions I_3 has a similar dependence, but under even only moderate absorp-tion conditions ($\mu t \simeq 1$) an important term in $\sin \alpha$ dominates I_3. This is sensitive to the sense of the phase difference, and examination of the con-trast of the first fringe from the exit surface enables the sense of α and hence the nature of the fault, whether extrinsic or intrinsic, to be determined. The magnitude of $\underline{R}$ can be determined from the reflections in which the fault is invisible. This occurs whenever $\alpha = 0$ or $2n\pi$ where n is integral (see, for example, Authier (1968) and Patel and Authier (1975)).

3.3.6. <u>Contrast of Stacking Faults in Traverse Topographs</u>

We again have two ways of generating the traverse topograph, by integrat-ing the section pattern along the trace of the intersection of the fault with the exit surface or calculating the contrast for a plane wave, and integrat-ing the intensities. The latter approach has a close parallel with the treatment in transmission electron microscopy and while this has not been exploited fully, Lang (1972) has analysed a stacking fault in quartz in a rather simplified manner which is instructive.

In a coordinate system defined in Fig. 3.16, where z is parallel to the Bragg planes and x is normal to them, the Bloch wave amplitudes above the fault for waves at the exact Bragg condition are proportional to $\exp(-i\pi gx) + \exp(i\pi gx)$. The first term comes from the $\underline{K}_g$ component and the second from the $\underline{K}_o$ component. On crossing the fault, the x coordinate transforms into

$$x' = x - \underline{g}.\underline{R}/|g|, \tag{3.32}$$

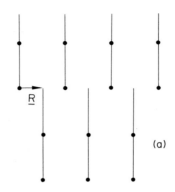

(a)

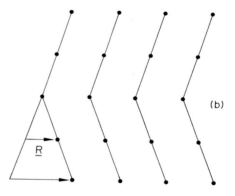

(b)

Fig. 3.17. (a) Displacement vector R corresponding to a stacking
         fault. (b) Displacement vector corresponding to a twin
         boundary.

and we can write each wavefield amplitude as proportional to
exp (-iπgx') exp(-iα/2) $\pm$ exp (iπgx') exp(iα/2), where α = 2πg.R.
   We can see quite clearly how the fault effects the Bloch wavefields by
introduction of a phase factor exp($\pm$iα/2) arising from the coordinate

transformation. The amplitudes of the wavefields below the fault can be determined by a straightforward process of identification. We note that in the special case illustrated in Fig. 3.17a, $\alpha = \pi$ and the branch 1 wavefields become the new branch 2 wavefields and vice versa. In the case considered by Lang, part of a fault lying close to the exit surface of a thick crystal is treated, so that above the fault only branch 1 wavefields at the exact Bragg condition are present. Defining the wavefields amplitudes below the fault as ψ_1 and ψ_2, we must have, due to amplitude matching at the boundary,

$$\psi_2\{\exp(-i\pi gx') + \exp(i\pi gx')\} + \psi_1\{\exp(-i\pi gx') - \exp(i\pi gx')\}$$

$$= \exp(-i\pi gx) - \exp(i\pi gx) = \exp(-i\pi gx')\,\exp(-i\alpha/2) -$$

$$\exp(i\pi gx')\,\exp(i\alpha/2). \tag{3.33}$$

Thus $\psi_2 = \cos \alpha/2$ and $\psi_1 = -i \sin \alpha/2$.

The diffracted wave amplitude at the exit surface can then be written, following section 1.5, as

$$D_g = -i \sin \alpha/2 - \cos \alpha \, \exp(2\pi iz/\xi_g), \tag{3.34}$$

where z is the distance from the exit face. This yields

$$I_g = 1 + \sin \alpha \, \sin 2\pi z/\xi_g \tag{3.35}$$

for the intensity.

Here we have the important term in $\sin \alpha$. Lang has formulated a rule for the contrast of the first fringe from the exit surface: "The contrast of the first fringe from the exit surface is positive if $\underline{g}.\underline{R}$ is positive".

While this analysis was successfully employed by Lang in quartz, Authier and Patel (1975) have pointed out that it is not always possible to perform such an anlysis on the <u>traverse</u> topograph. They have illustrated that in the case of a fault in silicon bounded by two partial dislocations, even when the first fringe of the section pattern is negative ($\underline{g}.\underline{R}<0$), the intense direct image swamps the contrast on integration over the trace of the fault with the exit surface resulting in positive contrast of the first fringe in the traverse topograph (Fig. 3.18). In section patterns there is no ambiguity and they urge topographers to use the section topograph for fault identification wherever possible. The simple theory outlined above can still be used to identify the first fringe in the centre of the section pattern.

In traverse topography the fringes close to the entrance surface tend to blur out as the integration is performed. In thick crystals, fringes corresponding to the fault in the centre of the crystal are not observed as only one wavefield, branch 1, reaches the exit surface. Good contrast fringes are only observed close to the exit surface where branch 1 wavefields above the fault excite both branch 1 and 2 wavefields below. Close to the exit surface appreciable absorption of the branch 2 wavefields does not occur and clear fringes result (Authier, 1968).

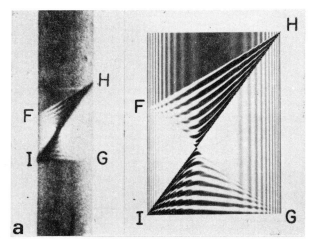

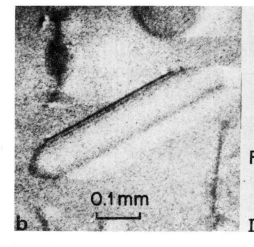

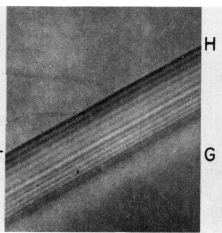

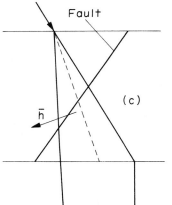

Fig. 3.18. (a) Experimental and computer simulated section topographs of an
inclined stacking fault in silicon.
The top fringe is light and the fault
is deduced to be extrinsic.
(b) Experimental and computer simulated traverse topograph of the same
fault. Note now that the top fault
is now dark (courtesy J.R. Patel).

3.3.7. Contrast of Twins

This may be treated in a manner similar to the stacking fault with a phase difference which varies with depth. However, this is equivalent to a change in the wavevectors $\underline{K}$, and this in turn corresponds to a shift in the tie points (at least in centrosymmetric structures). Inspection of Fig. 3.17b shows that this is exactly what one might expect from the geometry of a twin. The crystal below the boundary is misoriented with respect to that above, and we can find the amplitudes by considering the wavefields to decouple at the boundary and exciting new wavefields below it. As the Bragg planes are misoriented with respect to those above, the deviation parameter changes across the boundary and wavevector matching yields the new tie points.

In some crystals the misorientation is so great that diffraction occurs only in one twin, resulting in orientation contrast. Then thickness fringes may be observed across an inclined twin just as in a tapering edge of the crystal. The section topograph contrast of a lamella twin has been studied both experimentally and theoretically by Authier, Milne, and Sauvage (1968).

3.3.8. Contrast of Magnetic Domains

Contrast of magnetic domains in X-ray topographs arises from the magneto-strictive distortion, not through a direct interaction of the X-rays with the magnetization. In iron-silicon, for example, where the easy directions of magnetization are <100>, a tetragonal distortion occurs in the magnetization direction, with

$$c/a = 1 + 3\lambda/2 = 1 + 2 \times 10^{-5}, \tag{3.36}$$

where λ is the magnetostriction constant. The distortion is thus very small. Consideration of the magnetostatic energy shows that 90° domain walls should lie on {110} planes, and thus we see from Fig. 3.19 that the domain walls can be considered as coherent twinning boundaries. In an infinite crystal, 180° domain walls will not be visible in the topograph as no misorientation occurs between domains, and this is borne out experimentally. However, in certain diffracting conditions Lang has observed very weak contrast at 180° walls in iron-silicon which he interprets as an effect of the lattice relaxation where the domain wall intersects the surface. Weak dynamical contrast, which may be interpreted by the Penning-Polder theory, arises from this small distortion.

Polcarova and Gemperlova (1969) considered the distortion at a 90° domain wall and showed that it was identical to a coherent crystallographic twin provided that the wall thickness was neglected. As domain wall widths are usually of the order of hundreds of Ångstroms, this is negligible on the scale of the X-ray extinction distance.

Two types of contrast are obtained depending on the magnitude of the mag-netostriction constant λ. For large magnetostriction, e.g. in the anti-ferromagnet NiO, orientation contrast is observed. With a small beam diverg-ence only one set of domains reflects at once and the angular misorientation between domains can be measured from the rocking curve separation. Petroff and Mathiot (1974) exploited this using the Lang technique on terbium iron garnet at low temperature to measure the magnetostriction constants. Bradler and Polcarova (1972) made similar measurements on iron-silicon using the double crystal method. Magnetostriction constants derived by this method are in good agreement with those obtained by other techniques.

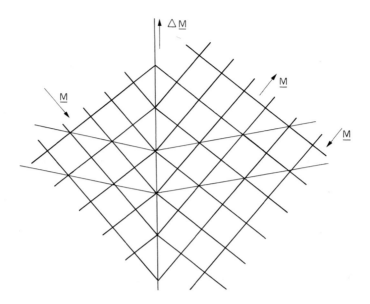

Fig. 3.19. Lattice misorientation across 90° and 180° walls in
          iron. Planes with $\underline{g}$ not perpendicular to $\Delta\underline{M}$ are seen
          to deviate on crossing the 90° wall. All planes are
          continuous across the 180° wall.

    For low magnetostriction both domains diffract simultaneously and no
orientation contrast is observed. However, provided that the misorientation
is greater than the perfect crystal reflecting range the wavefields will
decouple on crossing the domain wall and excite new wavefields in the second
domain. Although it is assumed that there is no phase change at the boundary itself (Polcarova, 1973), as we indicated in treating the twin, the new
tie points are at different positions on the dispersion surface from those
of the incident wavefields. Two types of contrast result. One is a net
change in intensity due to movement of the tie points and the other is an
interference effect from the newly created waves. Interference fringes can
then be observed across an inclined domain wall (Schlenker and Kleman, 1971).
However, care must be taken in the interpretation of such fringes, as
Polcarova and Lang (1971) observed fringes across walls in an (001) oriented
iron-silicon crystal which, unlike those in the (110) foil studied by
Schlenker and Kleman, did not change their spacing upon change of radiation.
Polcarova and Lang concluded that the contrast arose from the breaking up of
the 90° wall on {110} into zig zag fragments on the {111} planes and not due
to dynamical diffraction effects.
    The width of the contrast band is not usually related to the intrinsic
wall thickness. For example, in a (001) orientation iron-silicon foil a 90°
wall will be lying along $(1\bar{1}0)$ and the width of the image is dependent on the
tilt about the diffraction vector. As seen in Fig. 3.20, when $\gamma = 0$ for the
110 reflection, the image is a faint narrow line, the width of which is

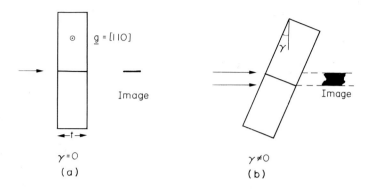

Fig. 3.20. (a) Image formed by a domain wall in an untilted crystal.
          (b) Image in a crystal tilted by an angle γ.

governed by the vertical dimensions of the source. When $\gamma \neq 0$, a broad band
appears, the width of which is t sin γ. The sense of the tilt of the latt-
ice planes across the boundary $\delta(\Delta\theta) = \pm 3\lambda_{100}$ sin γ is independent of the
sign of γ.
    In the early studies it was noted that the domain wall images disappeared
when (Polcarova and Kaczer, 1967)

$$(\underline{m}_1 - \underline{m}_2)\cdot\underline{g} = \Delta\underline{m}\cdot\underline{g} = 0. \tag{3.37}$$

This is in good agreement with the coherent twin model. Polcarova and Lang
(1968) noted that under moderate absorption conditions the net contrast of
domain walls appeared either bright or light depending on the sense of
$\Delta\underline{m}\cdot\underline{g}$. They developed a theory considering only the weakly absorbed wavefield and showed that the light-dark contrast was easily explicable in terms
of the shift in the tie point at the boundary. However, Polcarova (1969)
noted that the contrast was also a function of the exact absorption conditions, and she has recently developed a theory including both branches of
the dispersion surface (Polcarova, 1973).

3.3.9. Growth Bands

    Growth bands arise due to fluctuations in lattice parameter during crys-
tal growth. The lattice is distorted normal to the growth front by inclusion of impurities and, for example, a rapid fluctuation in temperature will
cause a large fluctuation in impurity content and hence lattice parameter.
When the change is large enough to cause interbranch scattering, fringes
analagous to those observed across stacking faults are observed (Zarka,
1969). As the distortion is normal to the growth front, the "fault"
disappears when the diffraction vector lies in the plane of the growth front.
That is, growth bands are invisible when

$$\underline{g}\cdot\underline{n} = 0, \tag{3.38}$$

where $\underline{n}$ is a vector parallel to the growth direction.

In addition to fringe contrast, very strong contrast occurs at the intersection of the growth band with the surface of the specimen. This surface
relaxation effect gives rise to a curving of the lattice planes at the surface of the crystal (Sauvage and Authier, 1965). Under moderate absorption
conditions black-white contrast, reversing with the diffraction vector, is
found, and this follows from the Penning-Polder theory. When the fluctuation in growth conditions is very large, long-range strain fields can
develop due to the impurity concentration, and these give rise to a direct
image of the strain at the fault.

3.3.10. Contrast on a Non-ideal Topograph

Finally, to cheer up those who are just beginning to take topographs,
Fig. 3.21 shows typical contrast from a poor quality crystal taken without

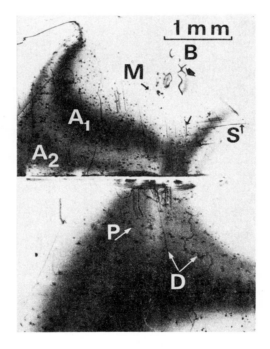

Fig. 3.21. Topograph of a bent crystal of $SnSe_2$. $11\bar{2}0$ reflection,
CuKα.

due care. Note particularly how the bend contours A_1 and A_2, corresponding to the $K\alpha_1$ and $K\alpha_2$ reflections, swirl across the micrograph. Large regions are white, as the crystal is too bent in those regions to diffract. Note that some bend contours B look similar to dislocations D. Further, attention is drawn to the superficial similarity between the images of dislocations D and scratches on the emulsion S, and between images of precipitates P and dirt on the emulsion M. The whole background is very grainy through inconsiderate development and the use of G5 emulsion rather than the recommended L4.

References

Aldred, P. and Hart, M. (1973) *Proc. Roy. Soc.* A332, 223, 239.
Ando, Y., Patel, J. R., and Kato, N. (1973) *J. Appl. Phys.* 44, 4405.
Authier, A. (1967) *Adv. X-ray Analysis (Plenum)* 10, 9.
Authier, A. (1968) *Phys. Stat. Sol.* 27, 77.
Authier, A. (1970) In *Modern Diffraction and Imaging Techniques*
       (ed. Amelinckx et al.), p. 481, North-Holland.
Authier, A., Malgrange, C., and Tournarie, M. (1968) *Acta Cryst.* A24, 126.
Authier, A., Milne, A. D., and Sauvage, M. (1968) *Phys. Stat. Sol.* 26, 469.
Authier, A. and Patel, J. R. (1975) *Phys. Stat. Sol. (a)* 27, 213.
Balibar, F. and Authier, A. (1967) *Phys. Stat. Sol.* 21, 413.
Balibar, F. and Malgrange, C. (1975) *Acta Cryst.* A31, 425.
Bradler, J. and Polcarova, M. (1972) *Phys. Stat. Sol. (a)* 9, 179.
Epelboin, Y. (1974) *J. Appl. Cryst.* 7, 372.
Epelboin, Y. and Lifchitz, A. (1974) *J. Appl. Cryst.* 7, 377.
Hart, M. (1963) *Ph.D. Thesis*, Bristol University.
Hart, M. and Milne, A.D. (1969) *Acta Cryst.* A25, 134.
Hart, M. and Milne, A.D. (1971) *Acta Cryst.* A31, 425.
Howie, A. and Basinski, Z. S. (1968) *Phil. Mag.* 17, 1039.
Kambe, K. (1963) *Z. Naturforschung* 189, 1010.
Kato, N. (1960) *Acta. Cryst.* 13, 349.
Kato, N. (1961) *Acta. Cryst.* 14, 526, 627.
Kato, N. (1963a) In *Crystallography and Crystal Perfection*
       (ed. Ramachandran), p. 153.
Kato, N. (1963b) *J. Phys. Soc. Japan* 18, 1785.
Kato, N. (1964) *J. Phys. Soc. Japan* 19, 67, 971.
Kato, N. (1969) *Acta Cryst.* A25, 119.
Kato, N. (1973) *Z. Naturforschung* 28a, 604.
Kato, N. (1974) In *Introduction to X-ray Diffraction* (ed. Azaroff),
       p. 425, McGraw-Hill.
Kato, N. and Lang, A. R. (1959) *Acta Cryst.* 12, 787.
Kato, N. and Patel, J. R. (1973) *J. Appl. Phys.* 44, 965.
Lang, A. R. (1970) In *Modern Diffraction and Imaging Techniques*
       (ed. Amelinckx), p. 407, North-Holland.
Lang, A. R. (1972) *Z. Naturforschung* 27a, 461.
Lang, A. R. and Polcarova, M. (1965) *Proc. Roy. Soc.* A285, 297.
Lawn, B. R. (1968) *Adv. X-ray Analysis (Plenum)* 11, 385.
Meieran, E. S. and Blech, I. A. (1965) *J. Appl. Phys.* 36, 3162.
Miltat, J. E. A. and Bowen, D. K. (1975) *J. Appl. Cryst.* 8, 657.
Patel, J. R. and Authier, A. (1975) *J. Appl. Phys.* 46, 118.
Patel, J. R. and Kato, N. (1973) *J. Appl. Phys.* 44, 971.
Penning, P. (1966) *Philips Res. Repts. Suppl.* No. 5.

Penning, P. and Goemans, A. H. (1968) *Phil. Mag.* <u>18</u>, 297.
Penning, P. and Polder, D. (1961) *Philips Res. Repts.* <u>16</u>, 419.
Petroff, J. F. and Mathiot, A. (1974) *Mat. Res. Bull.* <u>9</u>, 319.
Polcarova, M. (1969) *Trans. IEEE Magnetics* <u>MAG 5</u>. 536.
Polcarova, M. (1973) *Z. Naturforschung* <u>28a</u>, 639.
Polcarova, M. and Gemperlova, J. (1969) *Phys. Stat. Sol.* <u>32</u>, 769.
Polcarova, M. and Kaczer, J. (1967) *Phys. Stat. Sol.* <u>21</u>, 653.
Polcarova, M. and Lang, A. R. (1968) *Bull. Soc. Fr. Min. Crist.* <u>91</u>, 645.
Polcarova, M. and Lang, A. R. (1971) *Phys. Stat. Sol. (a)* <u>4</u>, 491.
Sauvage, M. and Authier, A. (1965) *Bull. Soc. Fr. Min. Crist.* <u>88</u>, 379.
Schlenker, M. and Kleman, M. (1971) *J. Phys.* <u>32</u>, C1, 156.
Shull, C. G. (1973) *J. Appl. Cryst.* <u>6</u>, 257.
Takagi, S. (1969) *J. Phys. Soc. Japan* <u>26</u>, 1239.
Tanner, B. K. (1972) *Phys. Stat. Sol. (a)* <u>10</u>, 381.
Tanner, B. K. (1973) *J. Appl. Cryst.* <u>6</u>, 31.
Taupin, D. (1964) *Bull. Soc. Fr. Min. Crist.* <u>87</u>, 469.
Zarka, A. (1969) *Bull. Soc. Fr. Min. Crist.* <u>92</u>, 160.

Appendix

Plane wave Pendellösung
Hashizume, H., Ishida, H., and Kohra, K. (1971) *Jap. J. Appl. Phys.* <u>10</u>, 514.

Energy flow
Gerward, L. (1973) *Z. Naturforschung* <u>28a</u>, 577.
Hashizume, H. and Kohra, K. (1970) *J. Phys. Soc. Japan* <u>29</u>, 805.

Spherical wave theory
Kato, N. (1964) *J. Phys. Soc. Japan* <u>19</u>, 971.
Kato, N. (1968) *J. Appl. Phys.* <u>39</u>, 2225, 2231.
Bedynska, T. (1971) *Phys. Stat. Sol. (a)* <u>4</u>, 627.

Eikonal theory review
Okkerse, B. and Penning, P. (1968) *Philips Tech. Rev.* <u>29</u>, 114.

Pendellösung fringes in distorted crystals
Hart, M. (1966) *Z. Physik* <u>189</u>, 269.
Fishman, Yu. M. and Lutsau, V. G. (1973) *Phys. Stat. Sol. (a)* <u>18</u>, 443.

Extensions and generalizations of the wave theory
Balibar, F. (1968) *Acta Cryst.* <u>A24</u>, 666.
Balibar, F. (1969) *Acta Cryst.* <u>A25</u>, 650.
Chukhovskii, F. N. (1974) *Soviet Phys. Cryst.* <u>19</u>
Chukhovskii, F. N. and Shtolberg, A. A. (1973) *Phys. Stat. Sol. (b)* <u>56</u>, K97.
Chukhovskii, F. N. and Shtolberg, A. A. (1970) *Phys. Stat. Sol.* <u>41</u>, 815.
Uragami, T. (1969) *J. Phys. Soc. Japan* <u>27</u>, 147.
Uragami, T. (1970) *J. Phys. Soc. Japan* <u>28</u>, 1508.
Uragami, T. (1971) *J. Phys. Soc. Japan* <u>31</u>, 1141.

Contrast of dislocations (Berg-Barrett)
Bedynska, T. (1973) *Phys. Stat. Sol. (a)* <u>18</u>, 147.
Roessler, B. and Armstrong, R. W. (1969) *Adv. X-ray Analysis (Plenum)* <u>12</u>, 139.

Contrast of dislocations (Lang)
Epelboin, Y. and Ribet, M. (1974) *Phys. Stat. Sol. (a)* <u>25</u>, 507.
Erofeev, V. N., Nikitenko, V. I., Polovinkina, V. I., and Suvorov, E. V.
        (1971) *Soviet Phys. Cryst.* <u>16</u>, 151.
Gerward, L. (1970) *Phys. Stat. Sol. (a)* <u>2</u>, 143.
Meieran, E. S. and Blech, I. A. (1972) *J. Appl. Phys.* <u>43</u>, 265.
Polovinkina, V. I., Suvorov, E. V., Chukhovskii, F. M., and Shtolberg, A. A.
        (1972) *Soviet Phys. Solid State* <u>14</u>,
Contrast of dislocations in interferometers
Christiansen, G., Gerward, L., and Lindegaard Anderson, A. (1971)
        J. Appl. Cryst. <u>4</u>, 370.
Fault contrast
Kato, N., Usami, K., and Katagawa, T. (1967) *Adv. X-ray Analysis (Plenum)*
        <u>10</u>, 46.
Authier, A. and Simon, D. (1968) *Acta Cryst.* <u>A24</u>, 517.
Simon, D. and Authier, A. (1968) *Acta Cryst.* <u>A24</u>, 427.
Domain Contrast
Schlenker, M., Brissoneau, P., and Perrier, J. P. (1968) *Bull. Soc. Fr. Min.
        Cryst.* <u>91</u>, 653.

Structure factor measurements
Hattori, H., Kuriyama, Y., and Kato, N. (1965) *J. Phys. Soc. Japan,* <u>20</u>, 988.
Batterman, B. W. and Patel, J. R. (1968) *J. Appl. Phys.* <u>39</u>, 1882.
Yamamoto, K., Homma, S., and Kato, N. (1968) *Acta Cryst.* <u>A24</u>, 312.
Bonse, U. and Hellkötter, H. (1969) *Z. Physik* <u>223</u>, 345.
Hart, M. and Milne, A. D. (1970) *Acta Cryst.* <u>A26</u>, 223.
Persson, E., Zielinska-Rohozinska, E., and Gerward, L. (1970) *Acta Cryst.*
        <u>A26</u>, 514.
Creagh, D. C. and Hart, M. (1970) *Phys. Stat. Sol.* <u>37</u>, 753.
Baker, J. F. C., Hart, M., and Hellior, J. (1973) *Z. Naturforschung* <u>28a</u>, 553.
Bonse, U. and Materlik, G. (1972) *Z Physik* <u>253</u>, 232.
Zuchkova, T.P., Polovinska, V. I., and Suvorov, E. V. (1975) *Soviet Phys.
        Solid State* <u>17</u>, 758.

CHAPTER 4

ANALYSIS OF CRYSTAL DEFECTS AND DISTORTIONS

The sensitivity to lattice distortions has been exploited in a number of ways, and many novel applications of X-ray topography have appeared in the last decade. Two extensive reviews of the applications of topography have been published, one by Lang (1970) and the other by Armstrong and Wu (1973). In this chapter we will examine only a few representative applications and attempt to see how much information can be extracted from crystals by the technique of X-ray topography.

4.1. DISLOCATIONS

4.1.1. Analysis of Burgers Vectors

Soon after the development of projection topography, Lang (1959) demonstrated its potential for Burgers vector analysis. The character of a dislocation is determined by its Burgers vector $\underline{b}$ and its line direction $\underline{u}$ (see, for example, Hull, 1975), and hence determination of $\underline{b}$ from topographs enables almost complete characterization. One can then see, for example, in which plane the dislocation will glide and hence relate this information to the macroscopic mechanical properties of the crystal. There is currently considerable interest in the early stages of plastic deformation, and this kind of information from X-ray topographs is an important contribution to our overall understanding of the processes involved.

In a distorted crystal, the phase difference introduced into the wave relative to an undistorted lattice is a product of the form $\underline{g}.\partial\underline{R}/\partial x_i$, where $\underline{R}$ is the atomic displacement around the defect.

Clearly when $\underline{g}.\underline{R} = 0$ no contrast is observed from the defect in the topographs. In a general isotropic elastic continuum the atomic displacement around a dislocation line in a plane normal to the line is

$$\underline{R}(r,\phi) = (2\pi)^{-1}\left\{\underline{b}\phi + \underline{b}_e \frac{\sin 2\phi}{4(1-\nu)} + \underline{b}\times\underline{u}\left|\frac{(1-2\nu)}{2(1-\nu)} \ln r + \frac{\cos 2\phi}{4(1-\nu)}\right|\right\},$$

(4.1)

where $\underline{b}_e$ is the edge component of Burgers vector and ν is Poisson's ratio. For a pure screw dislocation

$$\underline{R}(r,\phi) = \underline{b}\phi/2\pi,$$

(4.2)

and hence when

$$\underline{g} \cdot \underline{b} = 0 \qquad\qquad\qquad (4.3)$$

the dislocation is invisible in the topograph. The reason can be seen by
noting that for a screw dislocation there is an atomic displacement only in
the direction of the dislocation line. Thus lattice planes normal to the
dislocation are undeformed. As $\underline{b} \parallel \underline{u}$, this is equivalent to the condition
$\underline{g} \cdot \underline{b} = 0$. Similarly, it is easy to show that for a pure edge dislocation
this will be invisible when both $\underline{g} \cdot \underline{b} = 0$ and $g \cdot b \times u = 0$. A mixed orientation
dislocation, where $\underline{b}$ is neither parallel or perpendicular to $\underline{u}$ will never be
completely invisible. However, the second and third terms in (4.1) is usu-
ally much smaller than the first and approximate invisibility is often found
when $\underline{g} \cdot \underline{b} = 0$ for all types of dislocation.

   Here, then, is a way of analysing Burgers vectors. One takes a series of
topographs and observes in which reflections the dislocation is invisible.
Invisibility in two reflections with non-co-planar diffraction vectors is
required for an unambiguous determination of $\underline{b}$. The direction of $\underline{u}$ can be
determined either from stereo pairs or from the projected length of the dislocation in a number of reflections. This can be supplemented by section
topographs in which the position of the dislocation can be determined
precisely. Unfortunately, problems do arise. These include:

(a)   Bending of the specimen can make it extremely difficult to
      obtain good diffracting conditions at a particular dislocation
      in more than one reflection.

(b)   Contrast of other defects can obscure the dislocation under
      observation.

(c)   The specimen geometry may make it impossible to obtain
      reflections required.

As it is not always possible for the experimenter to control all these factors, sometimes complete Burgers vector analysis is precluded.

   The sense of the Burgers vector is much more difficult to determine.
Potentially the most general method is to compute the image of the dislocation in a section topograph and compare it with the experimental image. Both
sense and magnitude of Burgers vector markedly affect the section topograph
image and unambiguous identification of the sense, magnitude, and direction
of $\underline{b}$ is possible (Epelboin, 1974). It is the intermediary image which is
sensitive to the sense of Burgers vector and this is true also for the
traverse topograph. The intermediary image contributes an asymmetry to the
dislocation profile, and, by a careful analysis, Lang (1965a) was able to
determine the sense of an array of screw dislocations. Hart (1963) studied
the long range strain field using moderate absorption conditions. Under
these conditions, where interbranch scattering does not take place, the contrast far from the dislocation can be sensitive to the sense of lattice plane
curvature, and analysis using the Penning-Polder theory enables the sense of
$\underline{b}$ to be determined.

   Bonse (1958) has demonstrated that the double crystal method can be used
to determine the sense of $\underline{b}$ from a mapping of the lattice curvature, and a
similar principle was employed by Chikawa (1964a) in measurement of the shift
in the position of the rocking curve on either side of the dislocation.
Chikawa (1964b, 1965) also used an inclined slit to vary the angle of

incidence and determined the sense of $\underline{b}$ from the asymmetry of the images produced.

Analyses of Burgers vectors have been performed by a host of authors. A few studies are mentioned here either because they are, in my opinion, instructive or they contain high quality micrographs. Further examples are found in the appendix.

An early and spectacular study of a dislocation configuration was performed by Authier and Lang (1964) on a silicon crystal prepared by Dash. The specimen had been deformed by torsion and a single ended Frank-Read source had developed ten turns which were visible by X-ray topography. Stereo pairs were used to elucidate the three dimensional configuration and showed that in the vicinity of the spiral the spatial relationships were complex and that slip on secondary slip planes had occurred. For the pure screw dislocations, complete invisibility was found when $\underline{g}.\underline{b}$ = 0, but complete invisibility for 60° type dislocations was never found.

In a crystal of silicon deformed by thermal shock, Miltat and Christian (1973) showed that the obstacle at the end of a pile-up was a Lomer dislocation extending right through the crystal. Careful examination of the micrographs revealed that this dislocation did not act as a perfect barrier as the leading dislocations in the pile-up were seen to have passed the barrier by cross-slipping. The dislocation at the head of the pile-up was found to be in the process of crossing the barrier.

Lang and Polcarova (1965) carried out a study of dislocations in Fe-3.5% Si alloy and found a one-to-one correspondence between etch pits and dislocation outcrops. All Burgers vectors were parallel to <111> as expected in the b.c.c. structure. Dislocation images about 3 μm wide were recorded with AgKα radiation, in good agreement with the predictions of the simple mosaic model discussed in section 3.2.2 when the magnitude of the Burgers vector was $\frac{1}{2}$ <111>. In this paper the analysis is set out clearly step by step and it is a good teaching aid for the student of topography.

Similarly impressive was the study of Merlini and Young (1966) on the dislocations in annealed and lightly deformed copper crystals using anomalous transmission. Care must be taken in analysis of anomalous transmission topographs as in some crystals dislocations more than 100 microns or so from the exit surface give very diffuse images. For example, in silicon under even only moderate absorption conditions, confusion can arise if the exit and entrance surfaces are inadvertently interchanged. Some dislocations previously visible disappear and others appear, simply due to this spreading of the image. Merlini and Young used relatively thin, if highly absorbing, crystals and high-resolution topographs were consistently obtained. Dislocations of predominantly mixed orientation and Burgers vector parallel to <110> often did not lie on the {111} slip planes, and it was concluded that climb was important in establishing the equilibrium defect configuration. Image widths were not consistent with a simple mosaic model - as expected from their dynamical nature.

4.1.2. Study of the Early Stages of Plastic Deformation

During plastic deformation the dislocation density rises very rapidly, and thus X-ray topography is only applicable to studies of the pre-yield and very early stages of deformation if individual dislocations are to be resolved. Useful information can be obtained, particularly by the reflection Berg-Barrett technique, when the deformation is greater, but this concerns the slip band or sub-grain structure rather than the individual dislocations.

Although specimen preparation and handling problems can be enormous, several groups have performed fine high-resolution studies of plastic deformation.

Young and Sherrill used anomalous transmission topography to study dislocation motion under small applied stresses in copper. The specimens were strained in a spring-loaded tensile jig mounted directly on the Lang camera and topographs could be taken for incremental stress steps without relaxing the load. In recent studies they have shown how screw dislocations frequently cross-slipped in a double-slip orientation, filling the space between slip bands with dislocations (Young and Sherrill, 1971). The predominant type of dislocation was found to be a function of the specimen geometry rather than the Schmidt factor (Young and Sherrill, 1972). Two important points emerged from their studies. Firstly, that they were unable to determine either the nature of the sources or the generation mechanism of the dislocations in the slip bands, and, secondly, that grown-in dislocations moved very little and did not act as dislocation sources.

This latter observation contrasts sharply with the work of Rustad and Lohne (1971) on aluminium. In similar *in situ* studies they observed that grown-in dislocations did act as sources. They also observed the multiplication of dislocations close to the crystal surface and interpreted their results in terms of primary dislocations cross-slipping close to the surface and making anchor points for single ended Frank-Read sources (Lohne and Rustad, 1972). In an earlier study on aluminium in the temperature range 160-420°C, Nøst and Nes (1969) were able to establish by counting dislocations that the dislocation density N was related to the stress σ by

$$\sigma = 1.3GbN^{\frac{1}{2}} \qquad\qquad (4.4)$$

where G is the shear modulus.

Pegel and Becker (1969) found extensive helices in molybdenum which, on application of a small stress, collapsed by motion of the helical dislocations along their glide cylinders until only the screw components were left in the crystal. Measurements of the curvature of the specimen indicated that deformation was entirely by the motion of the helices. Estimates of the critical stress required to move the helices were in reasonable agreement with predictions from a theoretical model.

Sauvage (1968) observed dislocation reactions directly over a period of 10 months in a crystal of calcite and was able to show how the positions of the dislocations varied as a function of time. A splended ciné film was produced of these experiments. Recent developments in video techniques by Chikawa's group have enabled deformation studies to be performed dynamically rather than as step by step experiments. Dislocation movements over periods of the order of seconds can be recorded. In a high-temperature deformation experiment on silicon, Chikawa, Fujimoto, and Abe (1972) directly observed dislocation half loops being generated at the specimen surface and gliding across under an applied stress. The potential of the video systems, together with the use of synchrotron radiation for dynamic studies of plastic deformation, is considerable.

X-ray topography has been used in preference to etching techniques in studies of dislocation velocities, mainly at elevated temperatures in silicon. In a recent study (George *et al.*, 1972, 1973) dislocation half-loops were introduced into a silicon crystal by scratching and the mobility as a function of stress progressively measured. A known load was applied for a given time and the position of the dislocation determined from the topograph. As X-ray topography enables a determination of the Burgers vector to be performed, mobilities for screw and 60° dislocations could be obtained

unambiguously. They found that the activation energy for movement of 60°
dislocations was stress-dependent while the activation energy for movement
of screws was constant at high stresses and a function of stress at lower
values. All dislocation velocity studies on silicon reveal a mobility of
the form

$$v \propto (\sigma)^{m} \tag{4.5}$$

although the experimental results are not in agreement with current
theoretical predictions.

   Fukuda and Higashi (1973) have measured dislocation velocities in ice and
found a linear relationship between velocity and stress (m = 1). The X-ray
topographic data is at variance with the results of steady-state creep and
yield point experiments and it is clear that in the field of dislocation
mobility measurements there is much scope for further investigation.

4.1.3. Studies of Chemical Attack

   The effects of oxidation of metal surfaces have obvious technological
importance, and it is somewhat surprising that very few topographic studies
have been performed. Fiedler and Lang (1972) have observed the effect of
oxidation on a tin crystal over a period of several weeks. Beautiful arrays
of pure edge dislocations, seen running vertically in Fig. 4.1, were found to
grow parallel to the surface of the specimen at a rate of about 1 μm per
hour. The Burgers vector sense corresponded to a sheet of vacancies between
the dislocation core and the surface.
   Work on the effects of exposure of zinc to air has also indicated that
vacancy generation at the metal-oxide interface leads to dislocation climb
and an increase in dislocation density. Early work by Michell and Ogilvie
(1966) showed dislocation growth rates comparable with those found by
Fiedler and Lang. Recently, G'Sell and Champier (1975) have studied the
growth of dislocation loops and dipoles in zinc (Fig. 4.2). All dislocations climbed extensively, the Burgers vectors of the dipoles being such
that plastic deformation was eliminated as a mechanism of generation.
Comfortingly, when crystals were kept in an inert atmosphere or in vacuum,
defect evolution ceased. Technologically, the most important oxidation
process would seem to be that of iron, but this work has yet to be performed.

4.1.4. Device Control

   This is an area of great technological and economic importance and is
where X-ray topography has distinct advantages over other analytical techniques due to its large field of view and non-destructiveness.
Conveniently the substrate thickness used for device manufacture is ideal
for taking X-ray topographs. Much work is in progress, and if little
appears in the literature it is due to the obvious commercial interests
involved. Two types of device will be considered here - integrated circuit
semiconductor devices and magnetic bubble devices. X-ray topography has also
been employed fruitfully in directly imaging the modes of oscillation in
quartz transducers. The application of topography in this area has been
reviewed by Spencer (1968).
   The importance of dislocations in semiconductor devices was recognized
some 15 years ago when Prussin (1961) showed that dislocations could be

Fig. 4.1. Edge dislocation array generated by oxidation of the surface of a single crystal of tin. Field width 2.1 mm.
2$\bar{2}$0 reflection, AgKα radiation (courtesy A. R. Lang).

induced by excessive diffusion of impurity. Usually, in device manufacture,
windows are etched in an oxide film produced on the crystal surface by steam
oxidation, and subsequent diffusion of group III or V elements enables

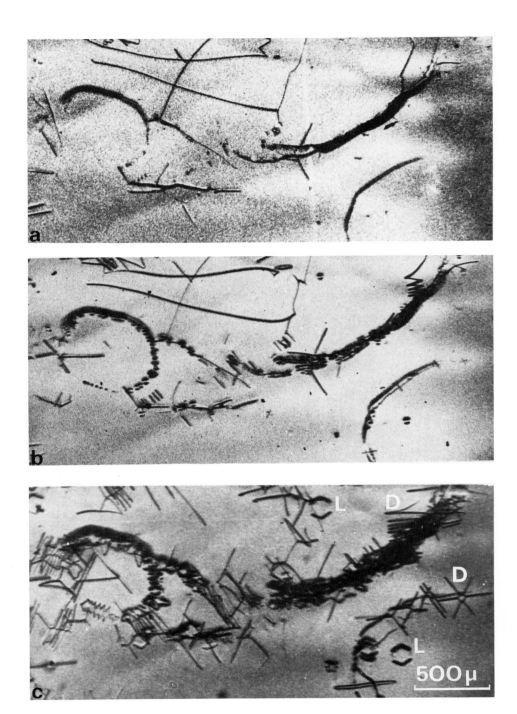

Fig. 4.2. Dislocation dipoles D and loops L in zinc formed by vacancy
           migration from the surface on oxidation. (a) After 1 day,
           (b) After 36 days, (c) After 109 days. 11$\bar{2}$0 reflection
           (courtesy C. G'Sell).

material of given carrier concentration to be produced in the localized
region of the window. Defects introduced in the processes, which are undertaken at high temperature, are potential hazards. Diffusion induced dislocations were first observed using X-ray topography by Schwuttke and Queisser
(1964). In silicon, they usually lie along <110> directions parallel to the
crystal surface and are of mixed or edge character. The creation of a dislocation enables the strain in the crystal caused by the mismatch in the
ionic radii of the matrix and dopant atoms to be relieved. The dislocations
propagate into the crystal by both glide and climb as processing is performed
at about 1000°C. At high dislocation densities they form a network having
the appearance in the X-ray topographs as illustrated in Fig. 4.3. There is

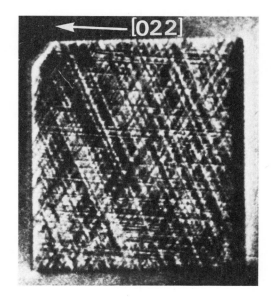

Fig. 4.3. Dislocation network inside the diffused region arising
          from the strains induced by the mismatch in ionic radii
          of the dopant (boron) and the matrix (silicon). 022
          reflection, CuKα radiation. Field width 1mm.

some debate as to whether the dislocations are confined to the diffused
region, but it is generally accepted that such diffusion induced dislocations
do not have a detrimental effect on device performance.

However, some dislocations do so, and Schwuttke and Fairfield (1966)
reported the observation of "emitter-edge" dislocations. These are dislocations generated beneath the edge of the oxide window used as a diffusion
mask and extend from the diffused to the undiffused region. An example of
emitter-edge dislocations can be found in Fig. 4.4. Schwuttke and co-workers
attribute their origin, not so much to the diffusion itself but to the

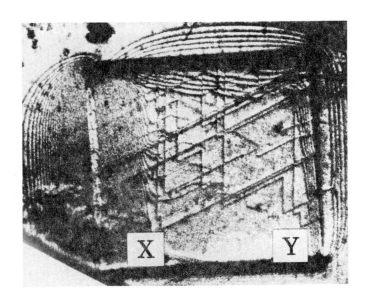

Fig. 4.4. Emitter-edge dislocation array at the boundary of the
           diffused region in a silicon crystal. The area inside
           the diamond shape has been heavily diffused with boron
           at 1250°C. The dislocation arrays X and Y extend from
           the diffused to the undiffused region. Note the inter-
           action between the emitter-edge dislocations and the
           straight dislocations inside the diffused region.
           220 reflection, MoKα radiation.

reversal of stress at the oxide edge during the diffusion. This effect has
been termed stress-jumping. Transistors containing emitter-edge dislocations
had consistently lower gain than those without such dislocations (Fairfield
and Schwuttke, 1968).
    Slip dislocations generated by plastic deformation during the high
temperature processing are extremely harmful. X-ray topographs superimposed
on device yield maps show a good correlation between soft diodes and dislocation slip bands. For example, Fairfield and Schwuttke (1966) found 90% of
the diodes in dislocation free areas were hard compared with only 20% in def-
ective regions. The practice of manufacturers at that time of scratching the
slice number on the back was shown to generate considerable numbers of dislocations. Similarly, Authier, Simon and Senes (1972) found a correlation

between dislocations and failure of Zener diodes.

Somewhat surprisingly, grown-in dislocations seem to have little effect on device yield, and Lawrence (1968) actually demonstrated that better yields were obtained in crystals containing a few grown-in dislocations. This was probably due to the fact that crystals at that time were of quite high oxygen content and the dislocations tended to soak up the impurity. Glaenzer and Jordan (1968) showed that edge dislocations generated at high temperatures were electrically inactive while those generated at lower temperatures were active, and this evidence is consistent with the topographic data on device yields.

Nowadays crystal growers can consistently grow dislocation-free crystals and production engineers avoid introducing dislocations by careful diffusion techniques, having learnt several lessons from the X-ray topographic studies. Figure 4.5 shows part of a typical topograph of an integrated circuit. The

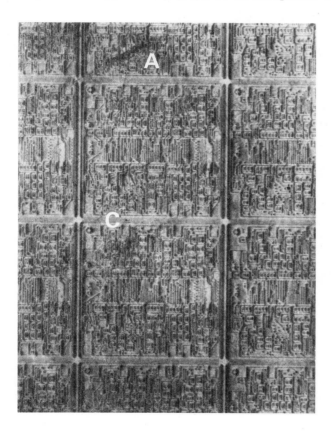

Fig. 4.5. Part of a topograph of a silicon wafer on which integrated circuits have been fabricated. Field of view 3.7 x 2.7 mm^2. MoKα radiation, 220 reflection.

components show up due to the strain in the lattice surrounding each diffused region. From such a complicated device, this intensity is strong and tends to mask dislocation contrast. At first sight, the region appears dislocation-free but on closer inspection, a section of a slip band at A is seen and some dislocations can be discerned in the area C. Figure 4.6 shows an example of larger components containing emitter-edge dislocations.

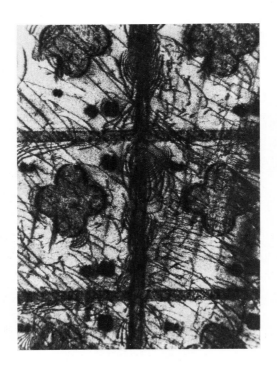

A useful review of the application of X-ray topography to semiconductor studies has been written by Meieran (1970), and Schwuttke (1970) also gives a good account of the state of knowledge at the beginning of this decade. While we may be approaching a situation where most of the device failures come from errors in mask alignment and production, X-ray topography has two important roles to play in the semiconductor industry. Firstly, as a routine inspection tool: in this it is important that the technique is rapid, though high resolution is not necessarily required. Chikawa and colleagues have applied their video system to on-line inspection and it would seem that there is a large potential for medium resolution video systems in this industrial field. Secondly, topography can be used effectively to monitor new manufacturing techniques. A good example is the use of ion implantation, where new defects were produced and X-ray topography has provided a valuable tool in their elimination (Schwuttke and Brack, 1969, 1973).

Fig. 4.6. Devices containing emitter-edge dislocations. Field width 4 mm. MoKα radiation, 220 reflection.

Dislocations, however, are not the only defects which impair device performance. Some years ago, Schwuttke (1970) and Goetzberger *et al.* (1963) showed that "swirl" patterns appearing in X-ray topographs were reproduced in the emission from an array of light emitting diodes (LEDs) fabricated on the slice. Nowadays device grade silicon crystals show no variation in contrast on Lang topographs (e.g. Fig. 2.18a). However, using the more sensitive (+ -) double crystal arrangement, diffuse bands of contrast are observed (Fig. 2.18b).

During crystal growth a cylindrically symmetric temperature distribution cannot be perfectly achieved and with every rotation of the seed a particular point on the crystal experiences a cyclic variation in temperature. This results in the impurity concentration in the crystal following a spiral

ramp with a pitch equal to the pulling rate divided by the rotation rate.
When a longitudinal section of a crystal is etched with CrO_3 and HF (the
Sirtl etch) shallow pits are revealed in a pattern of interrupted rings.
In a cross-section of the crystal these microdefects appear in a spiral
pattern. The defects are electrically active, and give rise to bright spots
arranged in a spiral pattern on the output monitor when a crystal of device
grade silicon is used to fabricate a silicon diode array vidicon.

De Kock and co-workers have developed a technique of lithium decoration
which enables the microdefects to be revealed in X-ray topographs. This
decoration is preferable to the older method of copper decoration as it is
not only faster but can be performed at a lower temperature. The Philips
group have observed two types of defect, the A type being larger than the
B type. The defects differ not only in size but in character and quenching
experiments revealed that B type clusters formed first, there being a well
defined temperature at which they turned into the large A type defects.
Moreover, the temperature at which the two types of cluster were formed was
found to be a function of growth rate. At growth rates above 5 mm/min nucleation of the defects was suppressed. It is believed that the defects are
vacancy clusters which require the formation of an oxygen-vacancy complex for
nucleation. The rate at which the complexes are formed is diffusion limited
and an increase in growth rate increases the cooling rate which gives less
time for the formation of the nuclei. Growth in an atmosphere of hydrogen
was also found to suppress cluster formation as the hydrogen reacted with the
oxygen and prevented the formation of the necessary oxygen-vacancy complex.
Unfortunately, precipitation of hydrogen occurred in the region close to the
crystal surface. Growth in pure argon at high growth rates is now used. An
excellent review of the work has been given by de Kock (1974) to which the
reader is referred for additional references.

As a final comment on silicon studies, it is worth noting that it is now
possible to produce crystals with such a low impurity content that even with
a double crystal arrangement sensitive to a few parts in 10^8 no contrast is
observed on the topographs. This very low oxygen content material has been
used for the fundamental lattice parameter measurements mentioned in section
2.11.

Double crystal topography may have been somewhat neglected in silicon
technology, but this has certainly not been the case in the field of magnetic devices. Magnetic bubbles are cylindrical magnetic domains which can
be produced in certain uniaxial magnetic materials under an applied field.
The cylindrical domains, whose axes lie normal to the specimen surface,
project into circles when viewed using the Faraday effect, and their resem-
blance with the two-dimensional soap-bubble raft of Bragg and Nye has led to
the coining of the term "magnetic bubbles". Under an in-plane magnetic field
the domains can be made to move across the specimen, and by evaporation of
permalloy tracks the domain movement can be controlled. (For details of
bubble devices and the material characteristics, see the recent book by O'Dell
(1975). Enough has been said to impress upon the reader that it is the
mobility of these domains which is exploited in bubble devices. For successful operation of, for example, a shift register, the movement of the bubble
must be predictable and controllable. As crystal defects interact and pin
domains due to the magnetostrictive interaction, sources of strain in the
crystal seriously affect device performance.

Currently, devices consist of a thin (about 5 μm) layer of a magnetic garnet grown epitaxially on a non-magnetic substrate, usually gadolinium gallium
garnet. Three types of defects are common - dislocations, faceted
regions, and growth striations. Two important types of dislocations occur
in the substrate - long straight dislocations running parallel to the growth

direction and large helical dislocations. Stacy, Pistorius, and Jannsen
(1974) have analysed the helical dislocations and concluded that they are
prismatic. The region inside the helix showed curious radial striations in
the topographs which were attributed to the existence of a smaller lattice
parameter inside the helix than outside. This is in agreement with the model
of loop formation by precipitation from the second phase proposed by Nes
(1973). Addition of excess Gd_2O_3 to the melt was found to suppress helix
formation (Matthews, Klokholm, and Plaskett, 1972). An example of such
helices viewed end-on is presented in Fig. 4.7. Use of X-ray topography in

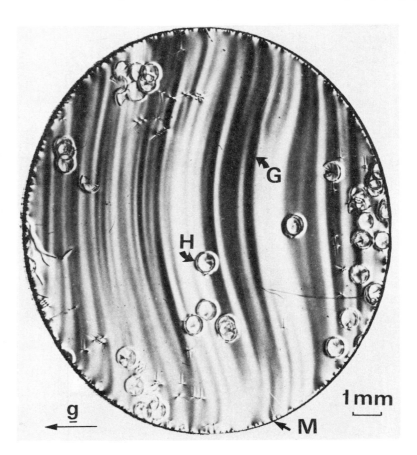

Fig. 4.7. Lang transmission topograph of a gadolinium gallium garnet
          crystal showing pronounced growth striations G and disloca-
          tion helices H. The helices are viewed end-on. Note also
          the mechanical damage M at the specimen edge. 800 reflec-
          tion, AgKα (courtesy A.D. Milne).

conjunction with optical birefringence microscopy has enabled conditions to be found where substrates can be grown free from dislocations using a technique of blocking the dislocations by facet formation in the early stages of growth (Cockayne and Roslington, 1973).

In faceted regions the mean lattice parameter is slightly smaller than in the unfaceted regions (Stacy, 1974), and hence these regions show up clearly in the X-ray topographs (Fig. 4.8). Luminescence studies indicate that the lattice parameter difference arises from trapping of oxygen vacancy complexes. Again, X-ray topography was employed in finding conditions where facet formation was suppressed (Basterfield, Prescott, and Cockayne, 1968). All three defects mentioned can propagate from the substrate into the epitaxial layer and hence impede bubble mobility. While it is possible to eliminate dislocations and facets, growth striae have not been totally eliminated. Although the lattice parameter fluctuation is small, sufficient stress can be present to affect device performance. Double crystal topography is ideal for the study of bubble devices as it is possible to image the substrate and film independently due to their different lattice parameter. The double crystal technique has been used quite extensively for measuring the difference in lattice parameter between substrate and film.

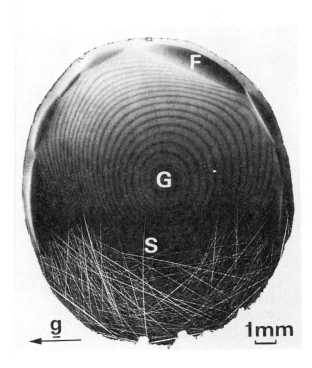

Fig. 4.8. Lang reflection topograph of a GGG crystal showing circular growth bands G and pronounced facets F. Scratches S give strong contrast. 888 reflection, AgKα (courtesy A.D. Milne).

Measurement of the angular separation of the peaks of the rocking curves enables the mismatch to be determined routinely to one part in 10^5 (Hart and Lloyd, 1975). The technique is particularly valuable for studies of multiple layers.

It has been found that with suitable specimen geometry, it is possible to prevent the lattice parameter fluctuations in the substrate associated with growth bands from being transmitted into the substrate. Strain amplitudes of less than 3×10^{-5} have been achieved (Stacy, 1974).

4.2.   PLANAR DEFECTS

    The general principles behind planar defect analysis have been given in
the last chapter. A fault plane introduces a phase shift

$$\alpha = 2\pi \underline{g}.\underline{R} \qquad\qquad\qquad (4.6)$$

into the X-ray wavefield. To determine $\underline{R}$ we must first establish in what
reflections the fault is invisible. Then $\underline{g}.\underline{R} = 0$ or n, where n is integral.
This determines the direction and magnitude of $\underline{R}$, and the contrast of the
first fringe from the exit surface must be examined to determine its sense.
(Preferably the examination should be of the section topograph.) Examples of
the detailed reasoning can be found in the papers by Kohra and Yoshimatsu
(1962) and Sauvage and Malgrange (1970).

4.2.1. Stacking Faults

    Stacking faults were first observed by X-ray topography in silicon by
Kohra and Yoshimatsu (1962) and have subsequently been studied by many other
workers. Kohra and Yoshimatsu verified the visibility criteria and studied
the fringe contrast from overlapping faults and in a tapering section.
Yoshimatsu (1964) extended the work and examined the bounding partials of
the faults. He also correlated the fault outcrops revealed by Dash's etchant
with the images on the X-ray topographs.
    Surely the most spectacular topographs are to be found in the study by
Meieran and Blech (1967) of stacking faults in annealed silicon web dendrite.
Although the presence of faults in as-grown web must be inferred from the
contrast of the bounding partials, upon annealing the partials dissociated
to distances up to a millimetre. The whole topograph (Fig. 4.9) is a checker-
board pattern of black and white faults which lie parallel to the specimen
surface. Intrinsic and extrinsic faults show opposite contrast in this
reflection.
    More recently, Patel and Authier (1975) have examined the contrast of dis-
location loops produced by heat-treating silicon. From the contrast of the
last fringe in the section topograph they were able to determine that
extrinsic faults were produced. Similar faults were found by Brummer and
Hofmann (1971) in gold-diffused silicon.
    Lawn, Kamiya, and Lang (1965) studied the contrast of stacking fault tet-
rahedra in the interior of natural diamond. The tetrahedra were single
faults bounded by Frank sessile dislocations. Weak ghost images observed in
the section topographs were interpreted using the spherical wave theory.
    Extensive faults have been observed in ice grown from a dilute solution
of ammonia (Oguro and Higashi, 1973). It appears that the presence of
ammonia leads to very large faults by the absorption of impurity at the rib-
bon fault between two partials, thus leading to stabilization of the fault.
Both Heidenreich-Shockley type faults with $\underline{R} = \frac{1}{3}<10\bar{1}0>$ and Shockley-Frank
faults with $\underline{R} = \frac{1}{6}<20\bar{2}3>$ were found. The faulted loops were observed to
shrink by a factor of about 3 over a period of 10 days, probably due to evaporation of the ammonia.
    On a rather more fundamental level, Sauvage and Malgrange (1970) examined
the fringes across a stacking fault in silicon produced under plane wave

Fig. 4.9. Extensive stacking faults lying parallel to the crystal
           surface in annealed silicon web dendrite. The black or
           white stacking faults S are bounded by partial disloca-
           tions. Field width 7 mm. 11$\bar{1}$ reflection (courtesy
           E.S. Meieran).

conditions. Across the plane wave image the deviation parameter varied and
the authors were able to show that the characteristic variation of fringe
spacing and profile predicted by Authier's (1968) theory was in good agreement with experiment. This study is particularly interesting as it has a

very close parallel with the electron microscopy situation.

4.2.2. Twins

The first X-ray topographic observation of planar defects was in quartz (Kato and Lang, 1959), and the elucidation of their nature has caused some trouble. Two common types of twin boundary are found in quartz—Brazil twins corresponding to a reflection in one of the {11$\bar{2}$0} planes normal to the a axis and Dauphiné twins corresponding to a rotation of π about the c axis. Both types of fault have been observed topographically (Lang, 1967). The fault vectors depend on the orientation of the boundary surfaces and analysis is complicated by the non-centrosymmetry of the crystal. Phakey (1969) performed a detailed analysis of Brazil twin fault vectors and observed stair-rod dislocations at the intersections of two faults. In a combined electron microscopy and X-ray topography study of Dauphiné twins, McLaren and Phakey (1968) noted fringe patterns even when the phase difference was zero. They explained these fringes in terms of the widely differing extinction distance in the two regions above and below the fault. A similar study of an inversion twin in BeO has been performed by Chikawa and Austermann (1968).

Contrast of twins appears in X-ray topographs either as extinction or orientation contrast. Across boundaries of parallel lattice twins, such as those considered above, extinction contrast is seen across the boundary in a similar way to which fringes are formed in stacking faults. In such twins, contrast can also be obtained from the body of the twin itself (as opposed to the boundary) due to the differing structure factor in the two twinned regions. Dauphiné twins in quartz give appreciably different diffracted intensities in some reflections. Lang (1965b) has also shown that Brazil twins can be distinguished by the different anomalous dispersion in the two regions.

Twins other than parallel lattice twins often are misoriented by substantial amounts and give rise to orientation contrast. In any one topograph only one twin will diffract and twin mapping has been applied to diamond (Lang, 1964), corundum (Wallace and White, 1967) and TbAℓO$_3$ (Wanklyn, Midgley, and Tanner, 1975) by this method.

In crystals containing twin lamellae and having a large misorientation, this effectively constitutes a non-diffracting lamella in the middle of the crystal. Contrast of such lamellae in calcite have been studied both experimentally and theoretically (Authier and Sauvage, 1966; Authier, Milne, and Sauvage, 1968).

4.2.3. Ferroelectric Domains

One of the most easily visualized forms of twin occurs as the ferroelectric domain, first observed topographically by Caslavsky and Polcarova (1964) in BaTiO$_3$. Below the transition temperature, barium titanate acquires a non-centrosymmetric tetragonal structure with the spontaneous polarization parallel to the c axis. 180° domain walls between directions of opposite polarization do not usually appear in the topographs as there is no long-range strain at the boundary and no misorientation between domains. They can, however, be revealed by anomalous dispersion which, by choice of suitable radiation, can be considerably different in the two domains. 180° domains have been seen in BaTiO$_3$ (Niizeki and Hasegawa, 1964) and Pb$_5$Ge$_3$O$_{11}$

(Sugii *et al.*, 1972) by this method.

90° domain walls between domains polarized parallel (*a*) and perpendicular (*c*) to the specimen surface are readily visible in barium titanate both by optical birefringence and X-ray topography. The wall is a twin boundary along {110} and as the difference $|c|-|a|$ is 1% the angular misorientation on crossing the wall is of the order of a degree. *a* and *c* domains do not diffract simultaneously in Lang topographs. As in the case of the ferro-magnetic domain, the wall becomes invisible when

$$\Delta\underline{p}\cdot\underline{g} = 0. \tag{4.7}$$

Planes corresponding to these reflections are continuous across the boundary (Polcarova, 1969b). Two examples of 90° ferroelectric domain wall contrast in $BaTiO_3$ are presented in Fig. 4.10. The crystal shown in Fig. 4.10a contains two sets of α domains running horizontally and vertically in the topograph. As a result of condition (4.7) the horizontal domain walls are invisible. The vertical domains show as thin lamellae of non-diffracting material. Figure 4.10b shows a highly perfect crystal of $BaTiO_3$ containing only one "V" shaped α domain. While condition (4.7) is almost obeyed, the domain is still observed to be visible here when $\Delta\underline{p}\cdot\underline{g} = 0$. However, the boundary is almost coherent as all regions of the crystal diffract simultaneously.

In the very important ferroelectric, triglycine sulphate (TGS), the role of the interaction of domain walls and lattice defects is somewhat uncertain. Malek *et al.* (1972) showed that as the crystal growth rate was reduced, so the dislocation density was reduced and the maximum permitivity rose. More significantly, they observed a direct correlation between the density of point defect clusters and the maximum permitivity, coercive field and Curie temperature in sections taken from a carefully grown crystal. The coercivity fell with decreasing defect density. The results show a significant inter-action between domain walls and defect clusters and the authors assign a lesser role to the dislocations.

Petroff (1969) observed two types of contrast associated with 180° domains in TGS. Strong contrast at the intersection of cylindrical domains with the surfaces perpendicular to the ferroelectric axis was interpreted as a space charge effect as it disappeared after substantial X-irradiation. Images observed both in topographs of plates cut parallel and perpendicular to the ferroelectric axis were ascribed to the distortion in the region of the wall itself. Assuming that the contrast arose solely from the strain in the wall, he calculated strain components on a theoretical model and deduced a value of 1500 Å for the wall thickness. This is very large compared with the width in barium titanate. While there is little doubt that the walls in TGS are very wide, recently some doubt has been cast on this interpretation of the contrast. Other observations on TGS (Takagi, Suzuki, and Watanabe, 1960; Takagi and Takahashi, 1975) indicate that the contrast is determined more by the imaginary part of the structure factor than the real part. Takagi's group suggest that the contrast arises from anomalous dispersion due to the different structure factor within the wall region rather than the strains.

Detailed domain studies have been made on sodium nitrite by Suzuki and Takagi (1966, 1967, 1971, 1972). Very thick 180° walls were also found in this material and the contrast is again attributed to the different structure factor within the wall.

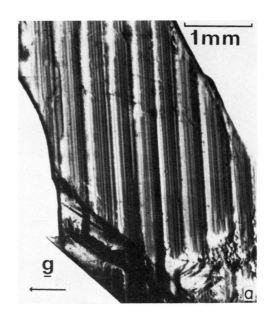

Fig. 4.10. Ferroelectric domains in $BaTiO_3$. Domains run parallel
to <100> directions. 200 reflections, MoKα radiation.

4.2.4. Magnetic Domains

X-ray topography has rapidly become an accepted tool for magnetic domain studies because of its unique ability of observing the magnetostriction directly. It also has an important advantage over most techniques for domain observation in being able to reveal interior domains in optically opaque materials (Polcarova, 1969a; Schlenker and Kleman, 1971). Its greatest potential lies, however, in the study of the coercivity where X-ray topography is capable of imaging both domain walls and lattice strains (e.g. dislocations) simultaneously. Up to the present, no major study of the interaction of a dislocation and a domain wall has been published. While correspondence between dislocation nodes and domain walls was found in iron-silicon by Wu and Roessler (1971) much more evidence is required before coincidence can be statistically eliminated.

There is evidence that dislocations do interact with domain walls. In an impressive experiment Kurtzig and Patel (1970) used X-ray topography only indirectly to show that movement of domain walls in a crystal of rare earth orthoferrite was impeded by the presence of dislocations. The dislocations were identified by X-ray topography and the domains observed by the Faraday effect.

Under an oscillatory magnetic field, the domain walls spent longer in the region of the dislocations than further away and hence a time exposure showed contrast around the region of the dislocations.

One feature only observable by X-ray topography is the strain which exists around the junction of a 180° wall with two 90° walls in iron-silicon The characteristic black-white "butterfly" images were studied qualitatively by Nagakura and Chikaura (1971) in iron whiskers. Recently, Miltat and Kleman (1973) have utilized the quasi-dislocation theory developed by Kleman and Schlenker (1972) to calculate the distortion at the so-called Y junction illustrated in Fig. 4.11(a). They found that the elastic field around the junction was characteristic of a wedge disclination. The surface relaxation determined the extent of the topographic contrast. Good agreement was found between theory and experimental observation. In particular, it was noted that junctions of opposite sign are frequently found in pairs (Fig. 4.11b) a feature arising from the large energy of the disclinations.

In some antiferromagnetic materials, X-ray topography is the only technique by which domains can be visualized. In classical antiferromagnets, domains cannot be imaged by the Bitter colloid, Faraday, or Kerr techniques because the net magnetization is zero. In some transparent materials, e.g. NiO, domains can be revealed by their birefringence but in some, e.g. $KNiF_3$, the birefringence is extremely low and domains have not been identified by optical techniques. The importance of X-ray topography in this area is clear. Antiferromagnetic domains were originally observed in NiO and CoO by Shimomura and co-workers (Yamada, Saito, and Shimomura, 1966; Saito, and Shimomura, 1961) and Kranjc (1969) measured the angular misorientation of domains in NiO by the Berg-Barrett Technique. Results were in good agreement with those predicted from the known values of the magnetostriction. Since these early experiments, Hosoya and Ando (1971) have used X-ray and neutron topography to observe spin density wave domains in chromium. Their results were extremely interesting, as the domain density was many orders of magnitude different from that predicted theoretically and this discrepancy remains a mystery. Recently, Schlenker, Baruchel, and Nouet (1973) and Safa, Midgley, and Tanner (1975) have studied domains in the important perovskites $KCoF_3$ and $KNiF_3$. An example of the domain structure found in $KNiF_3$ is given in Fig. 4.12. The sublattice magnetization direction is

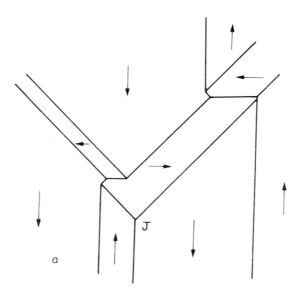

a

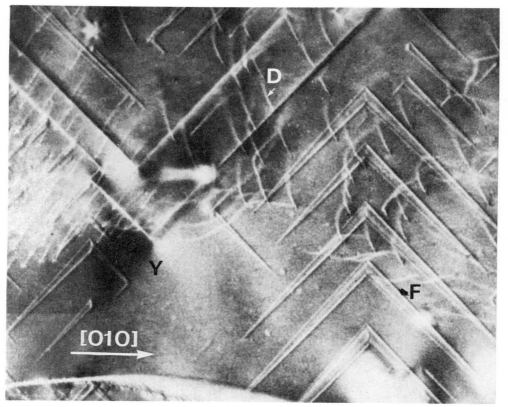

Fig. 4.11. (a) Schematic diagram of the "Y" junction between a
180° and two 90° domain walls in iron-silicon.
(b) Topograph of a "Y" junction in a (001) plate of
iron-silicon. Black-white "butterfly" contrast is seen
at the junction Y. Note also the 90° fir-tree closure
domain walls at F and the dynamical images of dislocations at D. Field width 1.4 mm, MoKα, 020 reflection
(courtesy J.E.A. Miltat).

Fig. 4.12. Antiferromagnetic domain structure in a flux-grown crystal of KNiF$_3$. Two sets of walls u and v are visible.

Both sets lie almost on {110} planes. 200 reflection,
AgKα (courtesy M. Safa).

<100> in both KCoF$_3$ and KNiF$_3$, and 90° domain walls lying on {110} planes
were found in both materials. When a small stress is applied to the crystal,
for example, by inconsiderate mounting, domains do not lie on exactly {110}
planes, and a "fir-tree"-like pattern very similar to that seen in iron can
be found.

4.2.5. Ion Implantation

Ion implantation, i.e. the forcing of impurity ions into a solid by bombardment with high energy particles, has important applications in the
manufacture of semiconductor devices. An interesting feature of the process
is that for high energy ions (about 2 MeV) an amorphous buried layer is
produced. This displaces the crystal lattice above and below resulting in a
translation fault. Bonse and Hart (1969) have analysed theoretically the
diffraction contrast from such a fault and showed that if the thickness of
the amorphous layer varied, equal thickness fringes would be expected in
asymmetric reflections. Experimentally, the intensity of the ion beam falls
off at the extrema and a lens-shaped amorphous layer is produced. Beautiful
fringes are, indeed, observed (Bonse, Hart, and Schwuttke, 1969). Using the
Bonse and Hart theory it is straight-forward to relate the visibility of the
fringes to the depth of the buried layer in the crystal. Measurements of
the depth of the layer from the fringe visibility were in good agreement with
the range of the ions in the solid and the depth of the layer measured by
sectioning. Subsequent work by Schwuttke and his colleagues has been concerned with the crystallographic and volume changes produced on annealing
(Schwuttke and Brack, 1969, 1973). It is worth noting that the production
of an amorphous layer is equivalent to the manufacture of a two component
monolithic interferometer.
    With low energy ions, a buried layer is not formed. The distortion in
the lattice following low energy ion implantation has been studied by
Gerward (1973) using both Lang topography and interferometry.
    Buried layers have been observed in silicon specimens following α particle
channelling investigations. 1 MeV α particles are quite extensively used to
obtain information concerning the sites of impurity atoms and it is not
widely appreciated how much damage can be introduced. Fig. 4.13 shows a
topograph of a silicon crystal following such an examination. In the areas
used for both 111 and 110 channels, marked A and B respectively, fringes
characteristic of a buried layer are observed and even in the alignment area
C, evidence of substantial damage is revealed.

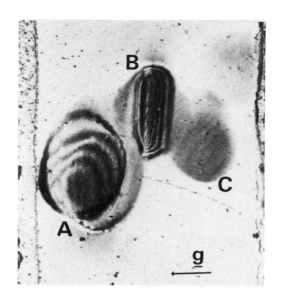

Fig. 4.13. Ion implantation damage in silicon following 1 MeV α
            particle backscattering analysis. 111 reflection, field
            width 2.5 mm.

References

Armstrong, R. W. and Wu, C. Cm. (1975) In *Microstructural Analysis, Tools and Techniques* (eds. McCall and Mueller), p. 169, Plenum.
Authier, A. (1968) *Phys. Stat. Sol.* 27, 77.
Authier, A. and Lang, A. R. (1964) *J. Appl. Phys.* 35,1956.
Authier, A., Milne, A. D., and Sauvage, M. (1968) *Phys. Stat. Sol.* 26, 469.
Authier, A. and Sauvage, M. (1966) *J. Phys.* 27, C3, 137.
Authier, A., Simon, D., and Senes, A. (1972) *Phys. Stat. Sol. (a)* 10, 233.
Basterfield, J., Prescott, M. J., and Cockayne, B. (1968) *J. Mater. Sci.* 3, 33.
Bonse, U. (1958) *Z. Physik* 153, 278.
Bonse, U. and Hart, M. (1969) *Phys. Stat. Sol.* 33, 351.
Bonse, U., Hart, M., and Scwuttke, G. H. (1969) *Phys. Stat. Sol.* 33, 361.
Brümmer, O. and Hofmann, M. (1971) *Phys. Stat. Sol. (a)* 5, 199.
Caslavsky, J. L. and Polcarova, M. (1964) *Czech. J. Phys.* B14, 454.
Chikawa, J-I. (1964a) *Appl. Phys. Lett.* 4, 159.
Chikawa, J-I. (1964b) *Appl. Phys. Lett.* 5, 31.
Chikawa, J-I. (1965) *J. Appl. Phys.* 36, 3496.
Chikawa, J-I. and Austermann, S. B. (1968) *J. Appl. Cryst.* 1, 165.
Chikawa, J-I., Fujimoto, I., and Abe, T. (1972) *Appl. Phys. Lett.* 21, 295.
Cockayne, B. and Roslington, J. M. (1973) *J. Mater, Sci.* 8, 601.
De Kock, A. J. R. (1974) *Philips Tech. Rev.* 34, 244.
Epelboin, Y. (1974) *J. Appl. Cryst.* 7, 372.
Fairfield, J. M. and Schwuttke, G. H. (1966) *J. Electrochem. Soc.* 113, 1229.
Fairfield, J. M. and Schwuttke, G. H. (1968) *J. Electrochem. Soc.* 115, 415.
Fiedler, R. and Lang, A. R. (1972) *J. Mater. Sci.* 7, 531.
Fukuda, A. and Higashi, A. (1973) *Crystal Lattice Defects* 4, 203.
George, A., Escaravage, C., Champier, G., and Schroter, W. (1972) *Phys. Stat. Sol. (b)* 53, 483.
George, A., Escaravage, C., Schroter, W., and Champier, G. (1973) *Crystal Lattice Defects* 4, 29.
Gerward, L. (1973) *Z Physik* 259, 313.
Glaenzer, R. H. and Jordan, C. G. (1968) *Phil. Mag.* 18, 717.
G'Sell, C. and Champier, G. (1975) *Phil. Mag.* 32, 283.
Goetzberger, A., McDonald, B., Haito, R., and Scarlett, R. M. (1963) *J. Appl. Phys.* 34, 1591.
Hart, M. (1963) *Ph.D. Thesis*, Bristol University.
Hart, M. and Lloyd, K. H. (1975) *J. Appl. Cryst.* 8, 42.
Hosoya, S. and Ando, M. (1971) *Phys. Rev. Lett.* 26, 321.
Hull, D. (1975) *Introduction to Dislocations*, Pergamon Press.
Kato, N. and Lang, A. R. (1959) *Acta Cryst.* 12, 787.
Kleman, M. and Schlenker, M. (1972) *J. Appl. Phys.* 43, 3184.
Kohra, K. and Yoshimatsu, M. (1972) *J. Phys. Soc. Japan* 17, 1041.
Kranjc, K. (1969) *J. Appl. Cryst.* 2, 227, 262.
Kurtzig, A. J. and Patel, J. R. (1970) *Phys. Lett.* 33A, 123.
Lang, A. R. (1959) *Acta Cryst.* 12, 249.
Lang, A. R. (1964) *Disc. Faraday Soc.* 38, 292.
Lang, A. R. (1965a) *Z. Naturforschung* 20a, 636.
Lang, A. R. (1965b) *Appl. Phys. Lett.* 7, 168.
Lang, A. R. (1967) *J. Phys. Chem. Solids* 28 suppl. 1, 833.
Lang, A. R. (1970) In *Modern Diffraction and Imaging Techniques in Materials Science* (ed. Amelinckx et al.), p. 407, North-Holland.
Lang, A.R. and Polcarova, M. (1965) *Proc. Roy. Soc.* A285, 297.
Lawn, B. R., Kamiaya, Y., and Lang, A. R. (1965) *Phil. Mag.* 12, 177.

Lawrence, J. E. (1968) *J. Electrochem. Soc.* 115, 860.
Lohne, O. and Rustad, O. (1972) *Phil. Mag.* 25, 529.
McLaren, A. C. and Phakey, P. P. (1968) *Phys. Stat. Sol.* 31, 723.
Malek, Z., Polcarova, M., Strajblova, J., and Janta, J. (1972) *Phys. Stat. Sol. (a)* 11, 195.
Matthews, J. W., Kokholm, E., and Plaskett, T. S. (1972) *AIP Conf. Proc.* 10, 271.
Meieran, E. S. (1970) *Siemens Review* 37, 39.
Meieran, E. S. and Blech, I. A. (1967) *J. Appl. Phys.* 38, 3495.
Merlini, A. and Young, F. W. Jr. (1966) *J. Phys.* 27, C3, 219.
Michell, D. and Ogilvie, G. J. (1966) *Phys. Stat. Sol.* 15. 83.
Miltat, J. E. A. and Christian, J. W. (1973) *Phil. Mag.* 27, 35.
Miltat, J. E. A. and Kleman, M. (1973) *Phil. Mag.* 28, 1015.
Nagakura, S. and Chikaura, Y. (1971) *J. Phys. Soc. Japan* 30, 495.
Nes, E. (1973) *Scripta Metall.* 7, 705.
Niizeki, N. and Hasegawa, M. (1964) *J. Phys. Soc. Japan* 19, 550.
Nøst, B. and Nes, E. (1969) *Acta. Metall.* 17, 13.
O'Dell, T. H. (1975) *Magnetic Bubbles*, Macmillan.
Oguro, M. and Higashi, A. (1973) In *Physics and Chemistry of Ice* (eds. Whalley, Jones, and Gold), p. 338, Roy. Soc. Canada.
Patel, J. R. and Authier, A. (1975) *J. Appl. Phys.* 46, 118.
Pegel, B. and Becker, C. (1969) *Phys. Stat. Sol.* 35, 157.
Petroff, J. F. (1969) *Phys. Stat. Sol.* 31, 285.
Phakey, P. P. (1969) *Phys. Stat. Sol.* 34, 105.
Polcarova, M. (1969a) *Trans. IEEE Magnetics* MAG 5, 536.
Polcarova, M. (1969b) *Czech. J. Phys.* B19, 657.
Prussin, S. (1961) *J. Appl. Phys.* 32, 1876.
Rustad, O. and Lohne, O. (1971) *Phys. Stat. Sol. (a)* 6, 153.
Safa, M., Midgley, D. and Tanner, B. K. (1975) *Phys. Stat. Sol. (a)* 28, K89.
Saito, S. and Shimomura, Y. (1961) *J. Phys. Soc. Japan* 16, 2351.
Sauvage, M. (1968) *Phys. Stat. Sol.* 29, 725.
Sauvage, M. and Malgrange, C. (1970) *Phys. Stat. Sol.* 37, 759.
Schlenker, M. and Kleman, M. (1971) *J. Phys.* 32 C1, 256.
Schlenker, M., Baruchel, J., and Nouet, J. (1973) *Proc. Int. Conf. on Magnetism*, Moscow, p. 368, IUPAP.
Schwuttke, G. H. (1970) *IBM Scientific Report* AFCRL-70-0110.
Schwuttke, G. H. and Brack, K. (1969) *Trans. AIME* 245, 475.
Schwuttke, G. H. and Brack, K. (1973) *Z. Naturforschung* 28a, 654.
Schwuttke, G. H. and Fairfield, J. M. (1966) *J. Appl. Phys.* 37, 4394.
Schwuttke, G. H. and Queisser, H. J. (1962) *J. Appl. Phys.* 33, 1540.
Spencer, W. J. (1968) In *Physical Accoustics* (ed. Mason) 5, Ch. 3, Academic Press.
Stacy, W. T. (1974) *J. Crystal Growth* 24/25, 137.
Stacy, W. T., Pistorius, J. A., and Jannsen, M. M. (1974) *J. Crystal Growth* 22, 37.
Sugii, K., Iwasaki, H., Itoh, Y., and Niizeki, N. (1972) *J. Crystal Growth* 16, 291.
Suzuki, S. and Takagi, M. (1966) *J. Phys. Soc. Japan* 21, 554.
Suzuki, S. and Takagi, M. (1967) *J. Phys. Soc. Japan* 23, 667.
Suzuki, S. and Takagi, M. (1971) *J. Phys. Soc. Japan* 30, 188.
Suzuki, S. and Takagi, M. (1972) *J. Phys. Soc. Japan* 32, 1302.
Takagi, M., Suzuki, S., and Watanabe, H. (1970) *J. Phys. Soc. Japan* 28, suppl. 369.
Takagi, M. and Takahashi, K. (1975) *Acta Cryst.* A31, S259.

Wallace, C. A. and White, E. A. D. (1967) *J. Phys. Chem. Solids* 28 suppl. 1, 431.
Wanklyn, B. M., Midgley, D., and Tanner, B. K. (1975) *J. Crystal Growth* 29, 281.
Wu, C. Cm. and Roessler, B. (1971) *Phys. Stat. Sol. (a)* 8, 571.
Yamada, T., Saito, S. and Shimomura, Y. (1966) *J. Phys. Soc. Japan* 21, 672.
Yoshimatsu, M. (1964) *Jap. J. Appl. Phys.* 3, 94.
Young, F. W. Jr. and Sherrill, F. A. (1971) *J. Appl. Phys.* 42, 230.
Young, F. W. Jr. and Sherrill, F. A. (1972) *J. Appl. Phys.* 43, 2949.

Appendix

Configuration Analysis
Dislocation reactions in silicon
Sauvage, M. and Simon, D. (1969) *Phys. Stat. Sol.* 35, 173.
Defects in oxidised silicon
Kawado, S. and Maruyama, T. (1972) *J. Appl. Cryst.* 5, 281.
Patel, J. R. (1972) *Bull. Soc. Fr. Min. Crist.* 95, 700.
Germanium
Tikhonov, L. V. and Khar'hova, G. V. (1968) *Soviet Phys. Cryst.* 12, 951.
Tikhonov, L. V. and Khar'hova, G. V. (1968) *Soviet Phys. Cryst.* 13, 386.
Wagatsuma, R. and Sumino, K. (1971) *J. Phys. Soc. Japan* 30, 891.
Wagatsuma, R., Kojima, K., and Sumino, K. (1971) *J. Appl. Phys.* 42, 867.
Indium antimonide
Sumino, K., Shimizu, H., and Wagatsuma, R. (1973) *J. Phys. Soc. Japan* 34, 1697.
Gallium phosphide
Kishini, S. (1974) *Jap. J. Appl. Phys.* 13, 587.
Indium phosphide and gallium phosphide
Clarke, R. C., Robertson, D. S., and Vere, A. W. (1973) *J. Mater. Sci.* 8, 1349.
Gadolinium molybdate
Kashiwada, Y. and Kishino, S. (1974) *Jap. J. Appl. Phys.* 13, 223.
Selenium
Naukkarinen, K. (1972) *Phys. Stat. Sol. (a)* 13, 399.
Arsenic
Shetty, M. N., Taylor, J. B., and Calvert, L. D. (1969) *Adv. X-ray Analysis (Plenum)* 12, 151.
Zinc sulphide (giant screw dislocations)
Mardix, S., Lang, A. R., and Blech, I. A. (1971) *Phil. Mag.* 24, 683.
Potassium dideuterium phosphate
Farabaugh, E. N. (1974) *J. Appl. Phys.* 45, 1905.
β-brass
Michell, D., Morton, A. J., and Smith, A. P. (1971) *Phys. Stat. Sol. (a)* 5, 341.
Iron
Futagami, K. (1971) *Jap. J. Appl. Phys.* 10, 814.
Aluminium
Lang, A. R. and Meyrick, G. (1959) *Phil. Mag.* 4, 878.
Ice
Jones, S. J. and Gilra, N. K. (1973) *Phil. Mag.* 27, 457.
Itagaki, K. (1970) *Adv. X-ray Analysis (Plenum)* 13, 527.

Plastic Deformation
Copper
Young, F. W. Jr. and Sherrill, F. A. (1967) *Canadian J. Phys.* <u>45</u>, 757.
Wilkens, M. (1967) *Canadian J. Phys.* <u>45</u>, 567.
Young, F. W. Jr. (1968) In *Dislocation Dynamics*, p. 313, McGraw-Hill,
        New York.
Minari, F., Pichaud, B., and Capella, L. (1975) *Phil. Mag.* <u>31</u>, 275.
Copper whiskers
Nittono, O. (1971) *Jap. J. Appl. Phys.* <u>10</u>, 188.
Iron
Coulon, G., Lecoq, J., and Escaig, B. (1974) *J. Phys.* <u>35</u>, 557.
Kushnir, I. P. and Sidokhin, E. F. (1974) *Soviet Phys. Solid State* <u>16</u>, 907.
Germanium
Wagatsuma, R., Sumino, K., Uchida, W., and Yamamoto, S. (1971) *J. Appl. Phys.*
        <u>42</u>, 222.

Dislocation Velocity Measurements
Suzuki, T. and Kojima, H. (1966) *Acta Metall.* <u>14</u>, 913.
Kannan, V. C. and Washburn, J. (1970) *J. Appl. Phys.* <u>41</u>, 3589.
Oki, S. and Futagami, K. (1974) *Jap. J. Appl. Phys.* <u>13</u>, 605.

Oxidation Studies
Zinc
Burns, S. J. and Roessler, B. (1972) *Phys. Stat. Sol. (a)* <u>13</u>, K91.

Diffusion Induced Defects
Ice (HF)
Jones, S. J. and Gilra, N. K. (1972) *Appl. Phys. Lett.* <u>20</u>, 319.
Silicon
Blech, I. A., Meieran, E. S. and Sello, H. (1965) *Appl. Phys. Lett.* <u>7</u>, 176.
Nimura, H., Ito, N., Nakau, T., and Nakahara, O. (1968) *Jap. J. Appl. Phys.*
        <u>7</u>, 43.
Barson, F., Hess, M. S. and Roy, M. M. (1968) *J. Electrochem. Soc.* <u>115</u>, 70.
Yoshida, M., Arata, H. and Terunuma, Y. (1968) *Jap. J. Appl. Phys.* <u>7</u>, 209.
Yukimoto, Y. (1969) *Jap. J. Appl. Phys.* <u>8</u>, 568.
Yoshida, M. and Saito, K. (1970) *Jap. J. Appl. Phys.* <u>9</u>, 1217.
Tamura, M. and Sugita, Y. (1973) *J. Appl. Phys.* <u>44</u>, 3442.
Matsui, J. and Shiraki, H. (1976) *Jap. J. Appl. Phys.* <u>15</u>, 73.

Stress in Films on Silicon
Blech, I. A. and Meieran, E. S. (1966) *Appl. Phys. Lett.* <u>9</u>, 245.
Blech, I. A. and Meieran, E. S. (1967) *J. Appl. Phys.* <u>38</u>, 2913.
Patel, J. R. and Kato, N. (1968) *Appl. Phys. Lett.* <u>13</u>, 40.
Kato, N. and Patel, J. R. (1968) *Appl. Phys. Lett.* <u>13</u>, 42.
Schwuttke, G. H. and Howard, J. K. (1968) *J. Appl. Phys.* <u>39</u>, 1581.
Saccocio, E. J. (1970) *Appl. Phys. Lett.* <u>17</u>, 149.
Saccocio, E. J. (1971) *J. Appl. Phys.* <u>42</u>, 3619.
Gajda, W. J. (1971) *Phys. Stat. Sol. (a)* <u>5</u>, K143.

Thermally Induced Dislocations in Silicon
Rai-Choudhury, P. and Takei, W. J. (1964) *J. Appl. Phys.* <u>40</u>, 4980.
Gerward, L. (1970) *Phys. Stat. Sol. (a)* <u>2</u>, 979.
Miltat, J. E. A. and Bowen, D. K. (1971) *Phys. Stat. Sol. (a)* <u>3</u>, 431.

Quartz
Application of a d.c. field
Yamashita, S. and Kato, N. (1975) *J. Appl. Cryst.* <u>8</u>, 623.
Application of an a.c. field
Isherwood, B. J. and Wallace, C. A. (1975) *J. Phys. D* <u>8</u>, 1827.
Goodall, F. N. and Wallace, C. A. (1975) *J. Phys. D* <u>8</u>, 1843.

Ferromagnetic Domains
Iron-silicon
Polcarova, M. and Lang, A. R. (1962) *Appl. Phys. Lett.* <u>1</u>, 13.
Roessler, B., Kramer, J. J., and Kuriyama, M. (1963) *Phys. Stat. Sol.* <u>11</u>, 117.
Wu, C. Cm. and Roessler, B. (1971) *J. Appl. Phys.* <u>42</u>, 1814.
Labrune, M. and Kleman, M. (1973) *J. Phys.* <u>34</u>, 79.
Iron whiskers
Chikaura, Y., Mori, T., and Nagakura, S. (1973) *J. Phys. Soc. Japan* <u>35</u>, 404.
Hagedorn, W. and Mende, H. H. (1970) *Z. Angewandte Phys.* <u>30</u>, 568.
Galinski, H. and Mende, H. H. (1975) *Phys. Stat. Sol. (a)* <u>27</u>, 35, 347.
Mende, H. H. and Galinski, H. (1974) *Appl. Phys.* <u>5</u>, 211.
Zone melted iron
Yamashita, T. and Mihara, A. (1971) *Jap. J. Appl. Phys.* <u>10</u>, 1661.

Ferrimagnetic Domains
Cobalt-zinc ferrite
Merz, K. M. (1960) *J. Appl. Phys.* <u>31</u>, 147.
Thulium orthoferrite
Patel, J. R., Van Uitert, L. G., and Mathiot, A. (1973) *J. Appl. Phys.* <u>44</u>, 3763.
Garnets
Basterfield, J. and Prescott, M. J. (1967) *J. Appl. Phys.* <u>38</u>, 3190.
Patel, J. R., Jackson, K. A., and Dillon, J. F. (1968) *J. Appl. Phys.* <u>39</u>, 3767.
Stacy, W. T. and Enz, U. (1972) *Trans. IEEE Magnetics* **MAG8**, 268.

Antiferromagnetic Domains
Cobalt oxide
Saito, S., Nakahigashi, K., and Shimomura, Y. (1966) *J. Phys. Soc. Japan* <u>21</u>, 850.
Nickel oxide
Saito, S. (1962) *J. Phys. Soc. Japan* <u>17</u>, 1287.
Blech, I. A. and Meieran, E. S. (1966) *Phil Mag.* <u>14</u>, 275.
Shimomura, Y. and Nakahigashi, K. (1970) *J. Appl. Cryst.* <u>3</u>, 548.
Nakahigashi, K., Fukuoka, N., and Shimomura, Y. (1975) *J. Phys. Soc. Japan* <u>38</u>, 1634.

Ferroelectric Domains
Barium titanate
Authier, A. (1968) *Bull. Soc. Fr. Min. Crist.* <u>91</u>, 666.
Lithium niobate
Wallace, C. A. (1970) *J. Appl. Cryst.* <u>3</u>, 546.
Gadolinium molybdate
Malgrange, C. and Glogarova, M. (1972) *J. Phys.* <u>33</u>, C2, 159.
Capelle, B. and Malgrange, C. (1973) *Phys. Stat. Sol. (a)* <u>20</u>, K5.

Ion Implantation Damage

Authier, A. and Montenay-Garestier, M. T. (1965) In *Effects des Rayonnments sur les Semiconductors*, p.79, Dunod, Paris.

Gerward, L., Christiansen, G., and Lindegaard Anderson, A. (1972) *Phys. Lett.* 39A, 63.

Brack, K. and Schwuttke, G. H. (1971) *Phys. Stat. Sol. (a)* 5, 711.

teKaat, E. H. and Schwuttke, G. H. (1972) *Adv. X-ray Analysis (Plenum)* 15, 504.

Uspenskaya, G. I., Genkin, V. M., and Tetel'baum, D. I. (1973) *Soviet Phys. Cryst.* 18, 224.

Auleytner, J., Furmanik, Z., Godwod, K., and Krylow, J. (1973) *Acta Phys. Pol.* A43, 507.

Radiation Damage

Copper (neutrons)

Baldwin, T. O., Sherrill, F. A., and Young, F. W. Jr. (1968) *J. Appl. Phys.* 39, 1541.

Hulett, L. D., Baldwin, T. O., Crump, J. C., and Young, F. W. Jr. (1968) *J. Appl. Phys.* 39, 3945.

Baldwin, T. O. and Thomas, J. E. (1968) *J. Appl. Phys.* 39, 4391.

Silicon (electrons)

Carron, G. J. (1966) *Appl. Phys. Lett.* 9, 355.

Carron, G. J. and Walford, L. K. (1967) *J. Appl. Phys.* 38, 3949.

Walford, L. K. and Carron, G. J. (1968) *J. Appl. Phys.* 39, 5802.

Triglycine sulphate (X-rays)

Polcarova, M., Bradler, J., and Janta, J. (1970) *Phys. Stat. Sol. (a)* 2, K137.

CHAPTER 5

CRYSTALS GROWN FROM SOLUTION

5.1. GROWTH FROM AQUEOUS SOLUTION

Crystals have been grown from aqueous solution for several centuries but only recently have techniques been developed to assess their crystallographic perfection. Using recipes derived over the years, crystal growers have been able to produce large optically perfect single crystals of a variety of compounds. In the last decade it has transpired that these crystals can also be very nearly dislocation free. Prior to the mid 1960s, it had been assumed that crystals such as NaCℓ grown from solution were good examples of mosaic crystals, and this assumption was almost unquestioned with respect to organic crystals. However, the 1966 experiments of Duckettt and Lang (1973) on hexamethaline tetramine, those of Ikeno, Maruyama, and Kato (1968) on NaCℓ, and those of Emara, Lawn, and Lang (1969) on potash alum all revealed that very low dislocation density, highly perfect single crystals could be grown without the use of sophisticated apparatus. A study of the dislocations in α-oxalic acid dihydrate (Michell, Smith, and Sabine, 1969) revealed a major discrepancy between the observed dislocation density and the mosaic block size predicted by use of Zachariasen's extinction correction. The approximation of such crystals to imperfect crystals was clearly unsound.

The importance of employing X-ray topography in the examination of such crystals should now be apparent, but it is worth noting how ideally suited organic crystals are to X-ray studies. The low absorption coefficients enables crystals to be examined in the as-grown state and thus there is no risk of introducing dislocations during any cutting or polishing operations.

In a short review, Authier (1972) has described the basic features revealed in the X-ray topographic studies of solution-grown crystals. We depict in Fig. 5.1 a typical defect configuration shown in a projection topograph. The dotted line represents the intersection of the growth sectors (i.e. the boundary between regions having grown on a common face), and we suppose here that the crystal has grown predominantly on {111} faces. From the seed S bundles of dislocations D have run almost normal to the growth faces. At small inclusions of solution or impurity I, dislocations were sometimes nucleated (in pairs), and these have also run normal to the growth faces.

Once growth conditions had stabilized and strains arising at the seed, e.g. from handling damage, been relaxed by dislocation nucleation, new dislocations were not generated. Thus as the crystal grew the dislocation density decreased. More importantly, the tendency for dislocations to bunch in specific directions has led to large regions becoming completely free from dislocations. In such regions spectacular dynamical diffraction effects may be observed, e.g. the Pendellösung fringes in Figs. 5.2 and 5.3.

The analysis of fringe patterns is however far from simple as exemplified

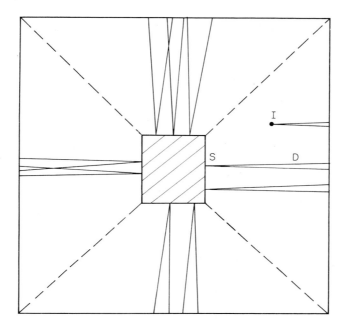

Fig. 5.1. Schematic representation of the dislocations found in a
          cubic crystal grown from aqueous solution. Dislocation
          lines D run normal to the growth faces.

in the first study of NaCl by Ikeno, Maruyama, and Kato, (1968). In addition
to identifiable Pendellösung fringes, two other types of fringe patterns
were observed. One set, which disobeyed Friedel's law (i.e. were of different contrast in $\pm$ g reflections) and which were invisible when the line of
the fringes was parallel to the diffraction vector, correlated optically
with growth steps on the bottom of the crystal. These "fringes" were
attributed to strains normal to the growth steps. The second fringe system
appeared only in the region where different growth sectors overlapped. Ikeno
et al. interpreted these as either double Pendellösung or Moiré fringes arising from a small distortion of 1 in 10^5 normal to the direction of growth.
In low extinction distance (strong) reflections, the two sectors are optically coherent and give rise to a Moiré system, whereas for long extinction
distance reflections (weak), the sectors diffract independently, giving rise
to a double Pendellösung system. However, the contrast and spacing of the
fringes was not well described by either model and it is likely that long-
range strains may have significantly disturbed the Pendellösung fringes.

    When growing NaCl, it is necessary to add Mn^{++} ions in order to grow good
quality crystals, and recently Kito and Kato (1974) have examined a similar
but pure system, $NaClO_3$. Dislocation bundles radiating from the seed were
found looking very similar to those in NaCl and those sketched in Fig. 5.1.

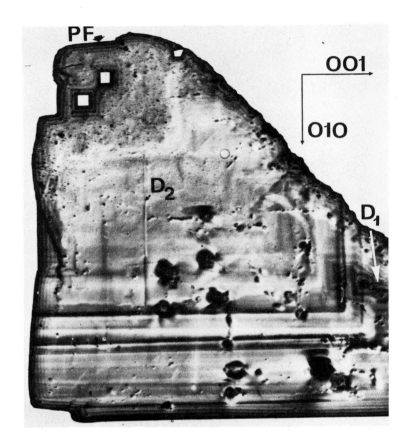

Fig. 5.2. Topograph of a flux-grown crystal of KNiF$_3$. Growth bands
         parallel to [001] and [010] indicate growth occurred on
         the cube faces at equal speed throughout growth.
         Dislocations D$_1$ and D$_2$ run normal to the growth faces.

         Thickness fringes PF indicate a high perfection. 011
         reflection. Field width 4mm. MoKα radiation (courtesy
         M. Safa).

The fringe system had very different properties to those in NaCl. In particular, fringes were found with opposite extinction rules to the Moiré patterns
recorded previously, i.e. the fringes disappeared when the diffraction vector
was normal to their direction. Kito and Kato interpret them in terms of a
rotation of the growth sectors with respect to one another by about 10^{-6} rad
resulting in a rotational Moiré or double Pendellösung pattern depending on
the extinction distance. It should be noted how important a role the section
topographs played in the interpretation of these fringes.

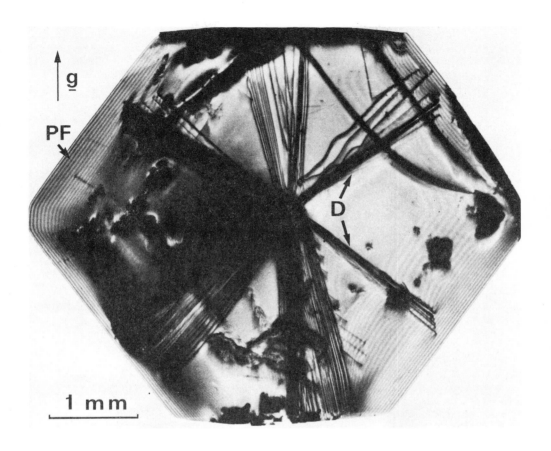

Fig. 5.3. Topograph of a crystal of (cubic) hexamethaline tetramine
grown from a solution of ethanol by slow isothermal evaporation. Note the clear Pendellösung fringes PF around
the edges of the crystal and the dislocations D running
normal to the growth faces in bundles (courtesy
A.R. Lang and R.A. Duckett).

When crystals are grown rapidly or without great care in eliminating
temperature or supersaturation fluctuations, growth bands develop. Growth
banding has become a common feature in X-ray topographs of solution grown
crystals and a good example can be found in the note on the perfection of
salol by Lefaucheux (1972). Figure 5.2 shows another example, this time from
high-temperature flux growth. The bands provide an excellent means of
determining the past growth history of the crystal and are now being
exploited in mineralogical studies (see Chapter 6).

5.1.1. Dislocations in Solution-Grown Crystals

It has become apparent that the dislocation configurations observed in solution-grown crystals exhibit a remarkable uniformity irrespective of crystal structure, chemical composition, or bonding. Klapper has postulated that this can be explained if during growth a dislocation always runs in a direction such that the elastic line energy per unit growth length is minimized. Equivalently, this implies that a dislocation runs in a direction such that the force on it due to the surface is always zero. Klapper and his colleagues have calculated the line energy $E(\underline{b},\underline{\ell},c_{ij})$ dislocations in crystals using anisotropic elasticity theory. The energy may be written

$$E(\underline{b},\underline{\ell},c_{ij}) = K(\underline{b},\underline{\ell},c_{ij})\ b^2\ \ell n(R/r)/4\pi, \qquad (5.1)$$

where K is the energy factor, R and r the outer and inner cut off radii of the dislocation, $\underline{b}$ the Burgers vector, $\underline{\ell}$ the line direction and c_{ij} the elastic constants.

The energy per unit growth length is then

$$W(\underline{b},\underline{\ell},c_{ij},\underline{n}) = E(\underline{b},\underline{\ell},c_{ij})/\cos\alpha. \qquad (5.2)$$

where $\underline{n}$ is the normal to the growth face and $\alpha\ (\underline{n},\ \underline{\ell})$ is the angle between $\underline{n}$ and $\overline{\ell}$.

As other variations are small, it is sufficient to consider only the variation in the energy factor

$$K_\alpha = K(\underline{b},\underline{\ell},c_{ij})/\cos\alpha \qquad (5.3)$$

This has been evaluated for a large number of dislocations in a variety of materials such as benzil (Klapper, 1971, 1972a), thiourea (Klapper, 1972b), KDP (Klapper, Fishman, and Lutsau, 1974) and lithium formate monohydrate (Klapper, 1973), and detailed comparisons between theory and experiment show quite good agreement. On Klapper's model, dislocations are expected to lie along directions corresponding to the minima in K_α, and in two limiting cases we can see the orientation by inspection of (5.3). Firstly, in an isotropic solid, K_α is independent of direction and the minimum in K_α occurs when $\alpha = 0$. This is equivalent to the dislocation running normal to the growth front as illustrated in Figs. 5.1 and 5.3. There is no dependence on Burgers vector, and the dislocations in NaCl, NaClO$_3$, and HMT are excellent examples. Secondly, when the anisotropy is large, very large angular deviations from the growth direction are found (up to 45° in ammonium hydrogen oxalate). This is in keeping with the observation that in organic crystals dislocations run straight but do not generally lie along crystallographic directions. We can also understand why a dislocation is refracted, sometimes up to 30° (Fig. 5.4), on crossing from one growth sector to another. The strong dependence of K_α on the angle α is sufficient in many cases to change the dislocation line direction, the loss of energy offsetting the increase due to the bend.

Generally, to make quantitative predictions, the energy factor must be evaluated numerically. When this is done, the agreement between theory and experiment is not total due to an important factor omitted from the theory -

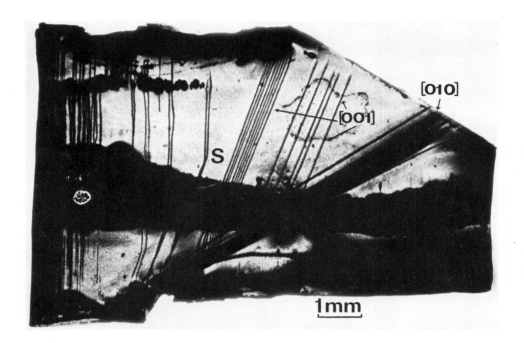

Fig. 5.4. Dislocation configuration in a crystal of ammonium hydro-
          gen oxalate. Note particularly the abrupt refraction of
          the dislocation as it crosses from one growth sector to
          another at S. In anisotropic AHO the dislocations make
          large angles to the growth face normals. Dislocations
          with b = [001] run in a different direction to those with
          b = [010] (courtesy H. Klapper).

the effect of the lattice structure. Klapper's theory assumes an elastic
continuum and this necessarily neglects the Peirels energy. In crystals such
as silicon and germanium it is well known that dislocations tend to lie along
<110> directions which are characterized by a low Peirels energy. Klapper
and Küppers (1973) noted that in ammonium hydrogen oxalate some dislocations
tended to follow the direction of minimum lattice energy rather than that of
minimum line energy. Calculations showed that in these cases the energy
factors for favoured lattice direction and the calculated preferential growth
direction were very similar.

    Provided the elastic constants are known, the direction of the dislocation
with respect to the growth face can be used to determine the Burgers vector.
If the elastic constants are not known, three rules apply.

    1. For a pure screw dislocation, $\ell \parallel n$.

2. For a pure edge dislocations, provided $\underline{n}$ is parallel to a
    twofold symmetry axis, $\underline{\ell} \parallel \underline{n}$.

3. For a mixed dislocation, $\underline{\ell}$ lies between $\underline{n}$ and $\underline{b}$.

The theory appears to be particularly useful in interpreting X-ray topographs of solution grown crystals and is, in my opinion, one of the more
important developments of the decade.

With a fair understanding of the dislocation configurations one may turn
to the problem of reducing the number of dislocations. Clearly it is
advantageous to avoid unnecessary mechanical damage to the seed. It is even
more important to ensure chemical purity of the solution and keep a uniform,
slow growth rate to avoid nucleation of dislocations at inclusions. As we
have seen, dislocations nucleated at the seed tend to bunch and are, therefore, somewhat less detrimental than those nucleated during growth. The
growth of TGS (triglycine sulphate) by Izrael *et al.* (1972) is an excellent
example of how the dislocation density can be reduced by careful control of
composition and growth parameters.

5.2. HYDROTHERMAL GROWTH

The most important hydrothermally grown crystal is undoubtedly quartz on
account of its uses as a piezoelectric, and it is not surprising that it was
the first hydrothermal crystal to be studied by X-ray topography. It had
additional interest in that large, highly perfect single crystals occur in
nature, and a comparison of the defects in the natural and synthetic crystals
was pertinent both from a geological as well as physical point of view. The
first high resolution work was performed by Lang and Muiscov (1967) following
an earlier study by Spencer and Haruta (1966). It is hardly surprising that
the defects found are similar to those grown from aqueous solution.
Dislocations nucleated at the seed run outwards to the crystal faces, and
very few dislocations are nucleated once the growth front is far from the
seed. A beautiful example of the defects found in a Y-cut plate is shown in
Fig. 5.5. The growth sectors associated with the Y-cut plate are sketched in
Fig. 5.6 which shows the geometry of a typical synthetic quartz crystal and
the nomenclature associated with the various cuts.

Although the dislocations run in the general direction of the growth
normal, they do not run exactly parallel to it. Lang and Muiscov noted that
in an X-cut dislocations in the Z-growth sector made angles up to 12° with
the $\underline{c}$ axis and that only 15% of the dislocations had a screw component of
Burgers vector. In Y- and Z-cut plates from the same Z-growth sector, dislocations are found to cluster into cell walls and the cell walls themselves
give rise to dynamical diffraction contrast similar to stacking fault
fringes. These cell walls correspond to the "cobble" texture, the network of
hillocks separated by grooves found on the Z sector growth faces. It was
noted by Brown and Thomas (1960) that the cobble texture was better
developed when greater concentrations of impurity were present. McLaren,
Osborne, and Saunders (1971) found that the dislocation density correlated
with the concentration of hydrogen in the crystal. Recently, Homma, and
Iwata (1973) have used electron probe micro-analysis to show that impurities
in fact segregate to the cobble grooves. They found from secondary electron
emission images that aluminium and iron were concentrated at the grooves and
this supported the earlier identification by Lang and Muiscov of the cell
wall contrast as arising from the interbranch scattering produced by a thin

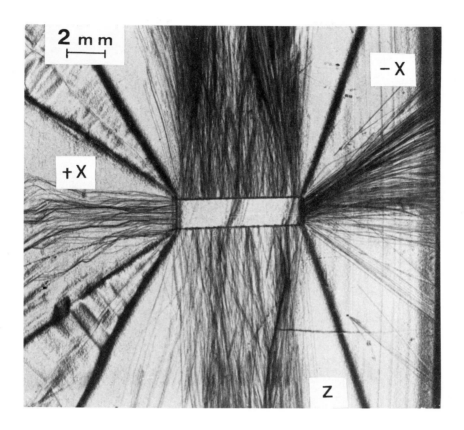

Fig. 5.5. Topograph of a Y-cut plate of hydrothermal quartz. Images
        of the growth sector boundaries are clearly visible.
        Virtually all the dislocations originate at the seed
        (courtesy M. Takagi).

strained layer of increased impurity content at the cell walls. Takagi,
Mineo, and Sato (1974) have observed peculiar images in section topographs
from the X growth sectors of a Y-cut plate which they also interpret as
images of the cell walls. McLaren *et al.* (1971) also found that the dislocations in the cell walls were predominantly of edge orientation and sugg-
ested that the cobbles arise because of a slower growth rate due to the high
concentrations of impurities at the walls.
    Lang and Muiscov observed that a strained surface 'skin' existed in the
crystals showing as enhanced diffraction contrast. This contrast disappeared
after a short etch in HF and was interpreted as arising from impurity
inclusion. Homma and Iwata confirmed by EPMA that this surface skin contained a higher concentration of sodium than the surrounding material. As we
shall see in the next section, such an impurity skin is commonly found in
flux grown crystals.

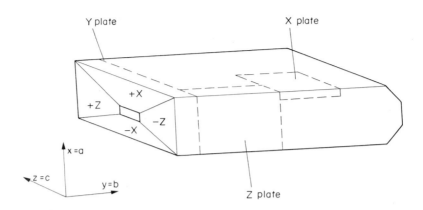

Fig. 5.6. Schematic diagram of the geometry of a synthetic quartz
          crystal grown on an X-cut seed.

Studies into the defect structure of hydrothermally grown calcite and the
effect of perturbations in the growth conditions have been performed by
Authier's group in Paris. Again, configurations typical of solution grown
crystals were found, and Epelboin, Zarka, and Klapper (1973) have used the
theory of minimum line energy to predict the angle the dislocations make
with the growth faces. Observed dislocation directions are in good agreement
with their predictions.

In an extremely informative study, Lefaucheux, Robert, and Authier (1973)
observed the effect of large perturbations on the growth conditions. They
examined the effects of abrupt changes of temperature and pressure on
spontaneously nucleated and seeded crystals. In crystals spontaneously
nucleated inside the autoclave, very few dislocations were observed
initially, the crystals being almost dislocation free. Upon reducing the
pressure in the autoclave abruptly by 25%, many dislocations were nucleated
at a growth band. These ran almost parallel to the growth face and had a
large screw component. The growth rate could be measured from the growth
banding on the topographs and it was found that the growth rate was increased
by nearly a factor of 5 when a strong screw component existed.

The other perturbation was arresting the growth for a period and again
dislocations were nucleated at the growth band corresponding to the
perturbation. However in both spontaneously nucleated and seeded crystals
when the dislocations did not have a strong screw component, no appreciable
change in growth rate was observed. On certain faces corresponding to
regions in the autoclave where the supersaturation was weak and constant
almost dislocation free growth took place. A most interesting feature of
this study was that the type of dislocation generated following the perturbations was very similar to those found in natural carbonates with the calcite
structure (Zarka, 1972).

Recently, Croxall et al. (1974) have investigated the perfection of
lithium doped zinc oxide crystals grown in a platinum lined autoclave.
Low-dislocation densities reminiscent of those in hydrothermally grown quartz

were found.

Belt (1967) in an earlier study of hydrothermally grown ruby found high densities of dislocations running normal to the growth face and substantial cracking. The configurations were remarkably similar to those found by Lefaucheux *et al.* after perturbation.

In summary then, we note that although very few studies of hydrothermally grown crystals have been published there exists a remarkable unity in the defect structures observed.

5.3.   FLUX GROWTH

While all crystal growth smacks somewhat of cookery, flux growth probably approaches alchemy more closely than any other branch of the subject. Flux growth is a form of high temperature solution growth, but because of the complex phase diagrams the kinetics are much more complicated than in hydrothermal growth. The main problems, apart from finding a crucible which will not be appreciably attacked by solvents such as PbO, PbF_2, and B_2O_3, is to avoid polycomponent growth. The skill in flux growth comes in obtaining just the correct conditions to nucleate and grow the crystals you require and suppress crystallization of other compositions. In this light it is not surprising that few perfection studies have been made on flux-grown crystals. Laudise (1970) makes no report of any perfection studies either on hydrothermal or flux grown crystals.

This neglect is unfortunate as the pioneering work of Austermann and co-workers in 1962 showed that flux-grown BeO were of very high perfection. Beryllium oxide, with its very low absorption coefficient made an excellent material for X-ray topographic study and a number of interesting defects were found in the as-grown crystals, of which the inversion twin has already been mentioned in section 4.2.2. Austermann, Newkirk, and Smith (1965) reported observations on BeO grown from a lithium molybdate flux. The crystals grew as prisms or pyramids extended in the c axis direction with the sharp tip as the nucleation site. Growth occurred on the basal plane at the blunt end. All crystals contained a twinned core which gave appreciably lower transmitted intensity than the surrounding crystals, and it is to be assumed that segregation of impurities had occurred. Beautiful axial screw dislocations or bundles of screw dislocations were observed outcropping at the basal plane growth face and presumably acting as the instruments of growth on that face. A projecting cone was frequently observed corresponding to the outcrop of the screw dislocations and, rarely, growth spirals were found where a large number of dislocations were present. In some crystals climb of the screw dislocation was noted, resulting in helical dislocations.

Subsequent topographic studies centred around magnetic garnets and barium titanate as there was a commercial interest in both materials as data storage media. As attempts were being made to exploit the properties of the ferromagnetic and ferroelectric domains, it is understandable that the topographic studies were concentrated on domain observations. Relatively little consideration was given to characterization of the as-grown crystals except to note that the perfection was sufficient to detect domain contrast in the topographs. In slices of flux-grown yttrium iron garnet both Patel, Jackson, and Dillon (1968) and Stacy and Enz (1972) found prominent growth bands but no dislocations. Belt (1969) did examine garnets from a defect point of view and found dislocation densities of between 10^2 and 10^4 cm^{-2}, the terminal growth sections being of the highest perfection. Unfortunately, by today's standards these crystals were highly imperfect confirming the suspicion that flux growth was not suitable for growth of low dislocation density crystals. Although Belt's (1967) study of ruby and sapphire showed areas up to 0.5 cm^2 dislocation free and clear Pendellösung fringes at tapering edges of the crystal, the specimens were rather warped and large area topographs were not obtained. Evidence for substantial flux inclusion came both from the strong contrast of the dislocations and from the fact that after annealing, many precipitates appeared.

The present author has examined several different kinds of flux-grown crystal in the as-grown state and found very low dislocation densities to be common in carefully grown crystals (Tanner, 1974). In iron borate, grown by spontaneous nucleation and slow cooling from a PbF_2 - PbO flux containing excess boric oxide, no dislocations were found and good images of the magnetic domains were observed in the topographs. Barium titanate is generally of lower perfection, but one crystal was found in a batch grown from a KF flux by the method of Remeika which contained only one ferroelectric $\underline{a}$ domain within the $\underline{c}$ domain plate and the only dislocations present were confined to this domain. Thick crystals of rare earth zircons (RVO_4 and $RAsO_4$) examined by anomalous transmission topography were found to be almost dislocation free in volume up to 15 mm^3. It is worth noting that crystals of mixed rare earth vanadates, crystals containing two rare earth ions were of similar perfection to the pure salts, indicating that when the ionic radii are very close, highly perfect alloy crystals can be grown. In none of these spontaneously nucleated crystals were dislocations observed originating at the nucleation site, a feature which seems to be quite common to spontaneously nucleated crystals in general.

Unfortunately, in all these studies, defect configurations typical of solution grown crystals were not observed and the overall unity of solution growth was not apparent. However, we have recently found crystals which contain such defect configurations. Figure 5.7 shows an anomalous transmission topograph of a crystal of $ErA\ell O_3$ grown from a PbO/PbF_2 flux containing MoO_3 as an additive. Growth occurred equally on two {100} faces and growth bands chart the growth history of the crystal. From a prominent growth band, presumably corresponding to a major perturbation in the growth conditions, two pairs of dislocations have been nucleated which run normal to the growth front (Wanklyn, Midgley, and Tanner, 1975).

Recent work on $KNiF_3$ grown from $PbCl_2$ flux has also revealed configurations typical of solution grown crystal. Figure 5.2 shows that highly perfect crystals can be grown, the thickness fringes at the tapered edges and etch pits providing clear evidence of the perfection. Dislocations D_1 and D_2 are pure edge in character and run normal to the growth faces, as expected from Klapper's theory.

Detailed studies have been made on the RVO_4 and RPO_4 systems grown by spontaneous nucleation and slow cooling from $Pb_2V_2O_7$ and $Pb_2P_2O_7$ fluxes respectively. (Tanner and Smith, 1975a, b). These studies are of particular interest as the crystals were necessarily recovered by pouring off the flux when it was molten. Normally the flux is allowed to solidify and the crystals are recovered by dissolving away the flux (leaching). In the case of RVO_4 and RPO_4 crystals unless a hot pouring technique is used, the crystals shatter under the differential thermal strains occurring on cooling. It was pleasing to find that hot pouring did not lead to catastrophic loss

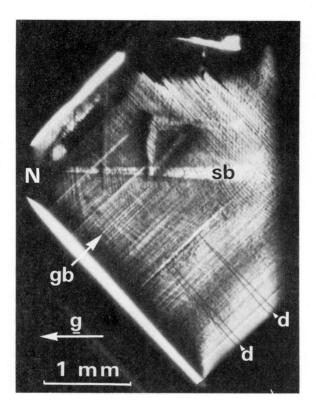

Fig. 5.7. Anomalous transmission topograph of a flux-grown crystal
of ErAℓO$_3$. Growth bands gb indicate equal velocity of
growth on the {100} faces. Two pairs of dislocations d
nucleated at the intense growth band and run normal to
the growth face. N is the nucleation point and sb is the
growth sector boundary. 110 reflection, MoKα radiation.

of perfection. In fact the only evidence of defects arising from the flux
pouring in the RVO$_4$ crystals is the existence of many precipitates close to
the surfaces of the crystals illustrated in Fig. 5.8. In the RPO$_4$ crystals
plastic deformation does take place, and although most dislocations are
associated with cracks some do seem to be nucleated at the specimen edge,
possibly from flux droplets remaining. A very recent study pf RAsO$_4$ has
again revealed large numbers of precipitates on the surface of the hot-
poured crystals whereas no precipitates are found when hot-pouring is not
performed. Otherwise there is little difference in the perfection of crystals which are hot-poured and those leached from the flux. Sometimes, however, crystals do show evidence of plastic deformation when hot pouring is
not undertaken.
    A feature clearly brought out in the study has been the existence of an

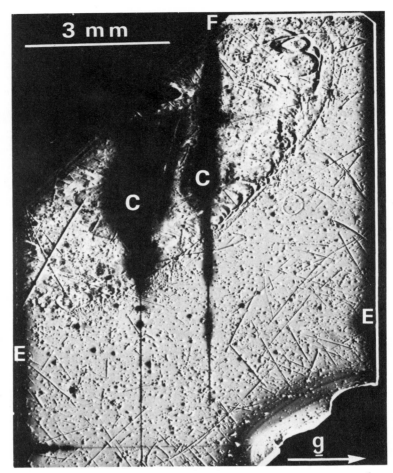

Fig. 5.8. Anomalous transmission topograph of an as-grown crystal
of $TmVO_4$. Note the many precipitates covering the
surfaces and also the cracks C. Also note the loss of
contrast at E due to the presence of the imperfect
"skin". 200 reflection, MoKα radiation. Crystal is
0.3 mm thick.

imperfect skin on flux grown crystals. This is similar to the skin found by
Lang and Muiscov (1967) in quartz and discussed earlier, and the associated
lattice curvature leads to loss of anomalous transmission at the crystal
edges E in Fig. 5.8. We note that when the distortion is normal to the
diffraction vector (e.g. at F) the skin contrast is not observed. We pres-
ume that in flux-grow crystals the impurity included in the lattice is the
flux. We noted the presence of the skin in a number of crystals (RVO_4 and
$RAlO_3$ being excellent examples) but never observed the skin effect in crystals containing many dislocations. The skin is significantly absent in RVO_4

crystals pulled from the flux (Garton, Smith, and Tanner, 1974) in which
dislocations nucleated at the seed run parallel to the growth direction.

An astonishing feature of the RVO_4 system is that large, optically visible flux inclusions could occur without propagation of dislocations into the
surrounding crystal. Close to inclusions, anomalous transmission is
observed, and dislocations do not seem to be generated by the inclusions.
This behaviour contrasts sharply with that of the BeO crystals examined by
Austerman et al. (1965) where bundles of dislocations were found to be
nucleated at flux inclusions. The answer may lie in the peculiar properties
of the flux used but is not understood at present.

What has become apparent in the last few years is that although it may
be extremely difficult to grow crystals from the flux due to the precipitation of other composition crystals, the resulting product can often be of
remarkably high perfection. Flux-grown crystals appear to behave as fairly
good examples of solution-grown crystals, and it is clear from current
activities that many more crystals will be characterized in the near future.

References

Austermann, S. B., Newkirk, J. B., and Smith, D. K. (1965) J. $Appl$. $Phys$.
        36, 3815.
Authier, A. (1972) J. $Crystal$ $Growth$ $13/14$, 34.
Belt, R. F. (1967) Adv. X-ray $Analysis$ $(Plenum)$ 10, 159.
Belt, R. F. (1969) J. $Appl$. $Phys$. 40, 1644.
Brown, C. S. and Thomas, L. A. (1960) J. $Phys$. $Chem$. $Solids$ 13, 337.
Croxall, D. F., Ward, R. C. C., Wallace, C. A., and Kell, R. C. (1974)
        J. $Crystal$ $Growth$ 22, 117.
Ducket, R. A. and Lang, A. R. (1973) J. $Crystal$ $Growth$ 18, 135.
Emara, S., Lawn, B. R., and Lang, A. R. (1969) $Phil$. Mag. 19, 157.
Epelboin, Y., Zarka, A., and Klapper, H. (1973) J. $Crystal$ $Growth$ 20, 103.
Garton, G., Smith, S. H., and Tanner, B. K. (1974) J. $Crystal$ $Growth$ 23, 335.
Homma, S. and Iwata, M. (1973) J. $Crystal$ $Growth$ 19, 125.
Ikeno, S., Maruyama, H., and Kato, N. (1968) J. $Crystal$ $Growth$ $3/4$, 683.
Izrael, A., Petroff, J. F., Authier, A., and Malek, Z. (1972) J. $Crystal$
        $Growth$ 16, 131.
Kito, I. and Kato, N. (1974) J. $Crystal$ $Growth$ $24/25$, 544.
Klapper, H. (1971) J. $Crystal$ $Growth$ 10, 13.
Klapper, H. (1972a) $Phys$. $Stat$. Sol. (a) 14, 99.
Klapper, H. (1972b) $Phys$. $Stat$. Sol. (a) 14, 443.
Klapper, H. (1973) Z. $Naturforschung$ $28a$, 614.
Klapper, H., Fishman, Yu. M., and Lutsau, V. G. (1974) $Phys$. $Stat$. Sol. (a)
        21, 115.
Klapper, H. and Küppers, H. (1973) $Acta$ $Cryst$. $A29$, 495.
Lang, A. R. (1967) J. $Phys$. $Chem$. $Solids$ 28 Suppl. 1, 833.
Lang, A. R. and Muiscov, V. F. (1967) J. $Appl$. $Phys$. 30, 2477.
Laudise, R. A. (1970) The $Growth$ of $Single$ $Crystals$, Prentice Hall.
Lefaucheux, F. (1972) J. $Crystal$ $Growth$ 16, 289.
Lefaucheux, F., Robert, M., and Authier, A. (1973) J. $Crystal$ $Growth$ 19, 329.
McLaren, A. C., Osborne, C. F., and Saunders, L. A. (1971) $Phys$. $Stat$. Sol.
        (a) 4, 235.
Michell, D., Smith, A. P., and Sabine, T. M. (1969) $Acta$ $Cryst$. $B25$, 2458.
Patel, J. R., Jackson, K. A. and Dillon, J. F. (1968) J. $Appl$. $Phys$. 39,
        3767.
Spencer, W. J. and Haruta, K. (1966) J. $Appl$. $Phys$. 37, 549.

Stacy, W. T. and Enz. U. (1972) *Trans. IEEE Magnetics* MAG8, 268.
Takagi, M., Mineo, H., and Sato, M. (1974) *J. Crystal Growth* 24/25, 541.
Tanner, B. K. (1974) *J. Crystal Growth* 24/25, 637.
Tanner, B. K. and Smith, S. H. (1975a) *J. Crystal Growth* 28, 77.
Tanner, B. K. and Smith, S. H. (1975b) *J. Crystal Growth* 30, 323.
Wanklyn, B. M. Midgley, D., and Tanner, B. K. (1975) *J. Crystal Growth*
        29, 281.
Zarka, A. (1972) *Bull Soc. Fr. Min. Crist.* 95, 24.

Appendix

Characterization studies of crystals grown from aqueous solution
Ammonium dihydrogen phosphate
Yoshimatsu, M. (1966) *Jap. J. Appl. Phys.* 5, 29.
Cyclotrimethylene trimitramine
McDermott, I. T. and Phakey, P.P. (1971) *Phys. Stat. Sol. (a)* 8, 505.
Hexamethylene tetramine
Di-Persio, J. and Escaig, B. (1972) *Crystal Lattice Defects* 3, 55.
Potassium dihydrogen phosphate
Fishman, Yu. M. (1972) *Soviet Phys. Cryst.* 17, 524.
Potassium dideuterium phosphate
Farabaugh, E. N. (1974) *J. Appl. Phys.* 45, 1905.
α-*sulphur*
Vergnoux, A. M., Riera, M., Ribet, J. L., and Ribet, M. (1971) *J. Crystal
        Growth* 10, 102.
Ribet, M. and Authier, A. (1972) *J. Crystal Growth* 16, 287.
Hampton, E. M., Hooper. R. M., Shah, B. S., Sherwood, J. N., Di-Persio, J.,
        and Escaig, B. (1974) *Phil. Mag.* 29, 743.
Triglycine sulphate
Authier, A. and Petroff, J. F. (1964) *C. R. Acad. Sci. Paris* 258, 4238.
Muiscov, V. F., Konstantinova, V. P., and Gusev, A. J. (1969) *Soviet Phys.
        Cryst.* 13, 791.

Characterization of gel-grown crystals
β-*silver iodide*
Caslavsky, J. L. and Suri, S. K. (1970) *J. Crystal Growth* 4, 213.

Characterization of flux-grown crystals
Sapphire
Takano, Y., Kohn, K., Kikuta, S., and Kohra, K. (1970) *Jap. J. Appl. Phys.*
        9, 847.
Magnesium aluminate spinel
Wang, C. C. and McFarlane, S. H. III (1968) *J. Crystal Growth* 3/4, 485.
Tabata, H., Okuda, H., and Ishii, E. (1973) *Jap. J. Appl. Phys.* 12, 7.
Isoberyl
Tabata, H., Ishii, E., and Okuda, H. (1974) *J. Crystal Growth* 24/25, 456.
Potassium nickel fluoride
Safa, M., Midgley, D., and Tanner, B. K. (1975) *Phys. Stat. Sol. (a)* 28, K89.
Rare earth othoferrites
Akaba, R. (1974) *J. Crystal Growth* 24/25, 537.
Titanium, zirconium, and hafnium biborides
Nakano, K., Hayashi, H., and Imura, T. (1974) *J. Crystal Growth* 24/25, 679.

CHAPTER **6**

NATURALLY OCCURRING CRYSTALS

6.1.   INTRODUCTION

It is understandable that there should be considerable interest in naturally occurring crystals. The rare beauty of natural gems has excited the mind of man for centuries. In recent years we have probed the structure of gemstones with a new collection of techniques and new beauties have been revealed. One of the techniques is, of course, X-ray topography, and historically we note that the pioneering work of Ramachandran and co-workers was performed on diamond. Soon after the development of the high resolution projection topographic technique, Lang (1959) turned his attention to natural quartz. It soon became clear that some gem quality crystals had an unexpectedly low dislocation density, and interest originally centred around defect analysis. However, X-ray topography has an important role to play in elucidating the growth history and as such may have important geological application. Particularly important in such studies are the growth bands or growth horizons which provide a definitive record of the past growth habit.

6.2.   DIAMOND

As diamonds were the first to be studied with X-ray topographic techniques, we will begin with them here. Lang has put much effort into studies of diamond by X-ray topography and it is a tribute to this work that X-ray topography is now an accepted tool for identification of genuine natural stones and detection of counterfeit ones. An excellent review of the work on diamonds has been written by Lang (1974a).

Diamonds usually grow on {111} faces as octahedra (known as normal growth) but occasionally growth on {100} faces is observed (known as abnormal or pathalogical growth). Due to dissolution of the points of the octahedron, normal diamonds often present a dodecahedral shape. Much discussion has taken place concerning the origin of this shape, and it was not until X-ray topography was applied that it became clear from the growth bands that growth had taken place on the {111} faces. Work on potash alum (Emara, Lawn, and Lang 1969) grown from aqueous solution showed that dissolution of octahedral alum crystals resulted in rounded dodecahedrons characteristic of diamonds. There is thus clear evidence that most diamonds have suffered considerable dissolution subsequent to growth.

Some normal diamonds have extremely low dislocation densities. From a central nucleus, a mere handful of dislocations radiate in non crystallographic directions (Fig. 6.1). Lang notes that these very low dislocation density crystals are generally tabular in habit and suggests that the absence of effective screw dislocation growth centres may inhibit the full development

Fig. 6.1. X-ray traverse topograph of the dislocation configuration
          in highly perfect natural diamond. 220 reflection, MoKα
          radiation. Crystal is 4 mm wide (courtesy A.R. Lang).

of the octahedra. Diamonds are excellent examples of crystals grown from
solution, the dislocation lines run very straight and occasionally a bundle
of dislocations is found to have been nucleated at an inclusion (Lang, 1963).
In diamonds suffering little dissolution, outcropping stacking faults have
been found (Lang, 1974a), and in rare specimens interior stacking fault tetrahedra have been identified (Lawn, Kamiya, and Lang, 1965). Normal diamonds
contain a relatively high nitrogen concentration which precipitates as platelets of Guinier-Preston zone type on {100} planes. These platelets have been
detected and analysed from the anomalous "spike" reflections produced by them
(Moore and Lang, 1972).

Abnormal diamonds are generally much less perfect. One variety, the
coated diamond, consists of a fibrous overgrowth on a reasonably perfect
core. Cracking of this core frequently accompanies coated growth. A very
sharp transition region is observed between normal growth and the coat. The
coat is highly imperfect and consists of fibres mutually misoriented by a
large fraction of a degree, (Kamiya and Lang, 1965). Growth is roughly parallel to <100> and impurity incorporation takes place unevenly. This type of
growth, looking superficially like the perturbed hydrothermally grown calcite
of Lefaucheux *et al.* and the hydrothermally grown ruby of Belt (see
Chapter 5) is not well understood. Further hydrothermal growth studies on
synthetic crystals would be valuable in this respect.

Two types of cubic diamond occur, one in which columnar growth occurs on
{111} planes with subsequent branching on other {111} planes to fill the
intervening space (Lang, 1974b) and the other in which growth occurs in
mixed habit on {100} and {111} faces. The beautiful octahedral "star" diamonds, on examination by X-ray topography reveal from the growth bands that

in the early stages of growth the crystals grew on the cubic faces. {111}
growth started to compete and eventually squeezed out the {100} growth resulting in final growth on {111} faces leaving an octahedral crystal
(Fig. 6.2). The {100} growth sectors appear quite perfect, and the X-ray

Fig. 6.2. Section topograph of an abnormal diamond. In this diamond
         growth occurred on both {100} and {111} faces, although the
         {111} growth finally determined the crystal morphology.
         Traces of the {100} planes are vertical and horizontal in
         the photograph. 440 reflection, MoKα radiation. Vertical
         height 5 mm (courtesy S. Suzuki).

intensity diffracted from them is much lower than the {111} sectors. This
suggests that {100} sectors are less rich in nitrogen than the {111} sectors
and that nitrogen is taken up preferentially on the {111} faces. This
influence of crystal habit on impurity content is a topic worth much further

investigation.

We should note that in his work, Lang has used a number of topographic techniques in the examination of one specimen. In addition to X-ray projection and section topography, he has employed ultra violet absorption topography, infra red absorption, optical birefringence microscopy, cathodoluminescence topography in addition to standard optical microscopy and etching. This is an excellent example of the fruitfulness of the use of complementary techniques, and it is hoped that the moral will not be lost on the reader. If one sees nothing by one technique alone, little can be concluded. However, if something is seen by one technique and nothing by a technique using a totally different imaging mode, one often finds a wealth of information available.

We cannot leave diamonds without reference to trigons, those peculiar triangular pits on the surfaces of natural diamonds which have been the subject of so much controversy. Lang (1964) showed a clear correlation between dislocation outcrops and naturally occurring trigons. The evidence seems overwhelmingly in favour of the suggestion that trigons are naturally occurring etch pits, resulting from the dissolution following growth.

6.3.   QUARTZ

Natural quartz crystals occur as hexagonal prisms capped usually by six facets. There are two types, the r and z faces corresponding to the major and minor rhombohedral sectors respectively. All are $\{1\bar{1}01\}$ type planes. As expected, X-ray topographs (Lang, 1967a, b) revealed growth bands parallel to $\{1\bar{1}01\}$ type planes indicating that growth had occurred on these planes throughout most of the crystal's growth history. Numerous examples were found of bundles of dislocations originating at inclusions and running approximately normal to the local growth front (Fig. 6.3). These dislocations are another good example of the nucleation of dislocations at foreign bodies in solution grown crystals. Quartz also provided the earliest example of a dislocation being refracted (see section 5.1.1.), i.e. changing direction on crossing from one growth sector to another (Lang, 1967a).

Much of the interest in natural quartz has centred around the nature of the two types of twin observed – the Brazil twin, visible optically, and the Dauphiné twin (see section 4.2.2.1.). X-ray observations showed that the Dauphiné twins were not planar, but stepped on a fine scale. Besides finding fringes at strong growth bands, Lang recorded fringe contrast at growth sector boundaries.

While optically perfect quartz yielded few surprises from a crystal growth standpoint, amethyst was more interesting. Amethysts contain laminated Brazil twins on the major rhombohedral growth sectors which all lie in the (0001) zone. When coloured, the colour bands run parallel to the twin lamellae and occur only in the r (major) growth sectors. Schlossin and Lang (1965) found that between the Brazil twins, and parallel to them lamellae of imperfect material occurred which deviated by less than a few degrees from the c axis. Use of X-ray absorption topography showed that the lamellae had a higher iron content than the surrounding crystal as well as containing a high density of dislocations. Superposition of optical, projection and absorption topographs demonstrated that the impurity lamellae lay between the Brazil twins. The twinned state appears to be a stable growth state and it was speculated that the dislocations in the imperfect region gave many growth centres allowing columnar growth rather than a spreading of growth sheets over the whole face.

Fig. 6.3. Traverse topograph of a natural quartz crystal. Growth
          bands delineate the growth faces and dislocations nuc-
          leated at inclusions run roughly normal to the growth
          bands. To the far right multiple overlapping Brazil
          twin boundaries are seen. $10\bar{1}0$ reflection, AgKα radia-
          tion. Field width 5.8 mm (courtesy A.R. Lang).

    Miuscov, Tzinober, and Gordienko (1973) have found a few natural quartz
crystals showing evidence of plastic deformation. Slip had occurred on the
rhombohedral basal prismatic planes and seemed to have occurred at high

temperature.

6.4. CALCITE, MAGNESITE, AND DOLOMITE

   Minerals with the calcite structure have been the object of detailed
study by Authier's group for some considerable time. Sauvage and Authier
studied natural calcite crystals from both Iceland and Mexico and observed
strong contrast from the growth bands where they intersected the crystal surface (Sauvage and Authier, 1965a). The early studies concentrated on the
twinning dislocations at the boundaries of twin lamellae (Sauvage and
Authier, 1965b), understanding the mechanism of twinning (Authier and
Sauvage, 1966), and studying the reactions between individual twinning
dislocations. The diffraction contrast produced by a lamella came in for
detailed analsis.
   The work of Zarka has been concerned mainly with the mineralogical aspects
of the research. He has made a detailed study of the defects in magnesite
and dolomite (Zarka, 1969), and found a good correlation between etch pits
and dislocation outcrops. Bundles of dislocations nucleated at precipitates
and running almost normal to the growth horizons were observed, the majority
of them being decorated ("dirty"). The growth bands, and hence growth,
occurred parallel to the natural faces of the rhombohedral crystal, i.e. on
the {010} planes. Recent calculations of the minimum elastic line energy of
dislocations have provided good agreement between predicted and observed
directions of the dislocation lines.

6.5. FLUORITE

   Beswick and Lang (1972) showed that in clear flawless specimens the dislocation density could be very low and confirmed the expected Burgers vectors as parallel to <110>. Again, bundles of dislocations were found running
normal to the growth front in good agreement with the calculations for minimum line energy (Epelboin, Zarka, and Klapper, 1973). In one crystal, however, they found sub-grain boundaries containing tangles of dislocations much
closer related to the configurations found in melt grown oxides such as MgO
than solution-grown crystals. It is presumed that this crystal suffered considerable plastic deformation subsequent to growth. The interpenetrating
twins found in Weardale fluorite appeared to trap impurities which led to the
presence of long-range strains.
   The inference that a departure from the cubic structure occurs due to
impurity incorporation during growth was confirmed by Tanner (1972). In a
{111} cleavage slice, growth bands were observed on the {100} growth planes
which gave particularly strong contrast at the intersection with the exit
surface. Corresponding to these growth sectors, strong growth sector birefringence was observed and the strongest growth bands were visible in the
birefringence micrographs. No diffraction contrast was observed at the
growth sector boundaries. Problems still remain in the interpretation of the
contrast of growth sector boundaries (Fishman and Lutsau, 1970; Parpia, 1975).
   Calas and Zarka (1973) correlated the coloured zones in a banded crystal
with the growth bands. As the colouration is believed to be caused by colour centres associated with heavy atom impurities, notably yttrium, it was
pleasing to find that the fluctuation in colour correlates with the fluctuation in lattice parameter, and hence, by inference, the impurity
concentration. A curious feature on one crystal, revealed from the growth

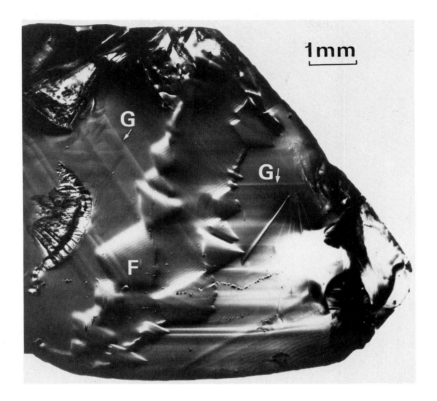

Fig. 6.4. Topograph of a {111} cleavage slice of a natural fluorite
          crystal. {100} growth faces are delineated by the major
          growth bands G. Note the successive attempts to grow on
          the slow {110} faces F (courtesy A. Zarka).

banding, was that it had tried, unsuccessfullly, several times to grow on the
slow (110) face. This left a zig-zag trace on the boundary between the (100)
and (010) growth sectors (Fig. 6.4). As in diamond this may be an impurity
effect.

6.6.   TOPAZ AND APATITE

    In topaz, as in fluorite, the directions of the dislocation lines were
found to be in excellent agreement with the predictions of Klapper's theory
(Epelboin *et al.*, 1973). Zarka (1974) has made an extensive study of the
Burgers vectors of the dislocations observed. Most dislocations originate
from impurity inclusions which are incorporated into the crystal during
growth. The bundles of dislocations run nearly normal to the growth front
and diverge slightly due to their elastic interaction with each other.
Refraction of the dislocations is observed across growth sector boundaries

and the change in direction is in good agreement with the theoretical
predictions.

Recently, Giacovazzo, Scandale, and Zarka (1975) have examined the growth
bands in sequential slices of crystals of Nigerian and Brazilian topaz and
were able to determine the growth history of the crystals. An independent,
and very similar, study of the growth history of a Japanese topaz crystal
has been published by Isogami and Sunagawa (1975). The crystal examined by
these workers was found to have changed its habit during growth. The use of
growth bands to chart the past morphology of minerals is certain to increase
in the future.

Phakey and Leonard (1970) studied natural apatite crystals and from the
strong growth banding verified that the hexagonal prisms grew on $\{10\bar{1}1\}$
surfaces. A very low dislocation density was observed, most dislocations
running normal to the growth faces. Most crystals contained a screw dislocation parallel to the long axis of the pyramid, the c axis.

6.7. BARITE, MICA AND ICE

Phakey and Aly (1970) made a study of natural barite plates and identified
the Burgers vectors of most of the dislocations observed. Relatively low
dislocation densities were observed but no information on the growth history
was obtained, as also was the case in the work on mica (Caslavsky and Vedam,
1970).

Finally we note that in 'natural crystals' we should include ice, although
the stability of these crystals is somewhat less than the other crystals
mentioned. Ice is quite common in certain parts of the world although large
undeformed single crystals are hard to acquire. Fukuda and Higashi (1969)
studied ice crystals cut from the Mendenhall Glacier in Alaska. High dislocation densities within sub-grains were found although the misorientation between these was less than 1 minute of arc. An interesting feature was that
the screw dislocation of Burgers vector $^1/_3[\bar{2}110]$ was much less common than
the other two equivalent types of dislocation. It can only be presumed that
this is a result of the plastic deformation occurring in the glacier.

6.8. RÉSUMÉ

The studies described above and in the previous chapter emphasise the
unity of the defect configurations occurring in solution grown crystals.
The presence of straight dislocations running in directions of minimum elastic energy is a dominating feature. In virtually all the systems tested so
far, Klapper's theory, in which the dislocation lines follow the direction of
minimum elastic energy, seems to predict the observed direction of dislocation lines very well. The abrupt refraction of the dislocations on crossing
growth sector boundaries is clearly understood on this model.

Growth rates on various faces have been found to be sensitive to the presence of screw dislocations, and definite evidence for screw dislocations
increasing the growth rate has been obtained. The final morphology is thus
determined both by the surface free energy and the dislocation configuration.
Further, it has been found that impurities can inhibit growth on otherwise
favourable faces. Screw dislocations do not seem to be a prerequisite for
appreciable growth rates.

Data on the defects found in solution grown crystals is now sufficiently
consistent that we may reasonably expect a satisfactory theory of solution

growth to emerge in the next few years.

References

Authier. A. and Sauvage, M. (1966) *J. Phys.* **27**, C3, 137.
Beswick, D. M. and Lang, A. R. (1972) *Phil. Mag.* **26**, 1057.
Calas, G. and Zarka, A. (1973) *Bull. Soc. Fr. Min. Crist.* **96**, 274.
Caslavsky, J. L. and Vedam, K. (1970) *Phil. Mag.* **22**, 255.
Emara, S., Lawn, B. R., and Lang, A. R. (1969) *Phil. Mag.* **19**, 157.
Epelboin, Y., Zarka, A., and Klapper, H. (1973) *J. Crystal Growth* **20**, 103.
Fishman, Yu. M. and Lutsau, V. G. (1970) *Phys. Stat. Sol. (a)* **3**, 829.
Fukuda, A. and Higashi, A. (1969) *Jap. J. Appl. Phys.* **8**, 993.
Giacovazzo, C., Scandale, E., and Zarka, A. (1975) *J. Appl. Cryst.* **8**, 315.
Isogami, M. and Sunagawa, I. (1975) *American Mineralogist* **60**, 889.
Kamiya, Y. and Lang, A. R. (1965) *Phil. Mag.* **11**, 347.
Lang, A. R. (1959) *Acta. Cryst.* **12**, 249.
Lang, A. R. (1963) *Brit. J. Appl. Phys.* **14**, 904.
Lang, A. R. (1964) *Proc. Roy. Soc.* **A278**, 234.
Lang, A. R. (1967a) *J. Phys. Chem. Solids* **28**, Suppl. 1, 833.
Lang, A. R. (1967b) *Adv. X-ray Analysis (Plenum)* **10**, 91.
Lang, A. R. (1974a) *J. Crystal Growth* **24/25**, 108.
Lang, A. R. (1974b) *J. Crystal Growth* **23**, 151.
Lawn, B. R., Kamiya, Y., and Lang, A. R. (1965) *Phil. Mag.* **12**, 177.
Lefaucheux, F., Robert, M. C., and Authier, A. (1973) *J. Crystal Growth* **19**, 329.
Miuscov, V. F., Tsinobar, L. I., and Gordienko, L. A. (1973) *Soviet Phys. Cryst.* **18**, 209.
Moore, M. and Lang, A. R. (1972) *Phil. Mag.* **25**, 219.
Parpia, D. Y. (1975) *J. Appl. Cryst.* **8**, 203.
Phakey, P.P. and Aly, A. M. (1970) *Phil. Mag.* **22**, 1217.
Phakey, P. P. and Leonard, J. R. (1970) *J. Appl. Cryst.* **3**, 38.
Sauvage, M. and Authier, A. (1965a) *Bull. Soc. Fr. Min. Crist.* **88**, 379.
Sauvage, M. and Authier, A. (1965b) *Phys. Stat. Sol.* **12**, K73.
Schlossin, H. H. and Lang, A. R. (1965) *Phil. Mag.* **12**, 283.
Tanner, B. K. (1972) *Phys. Stat. Sol. (a)* **14**, K9.
Zarka, A. (1969) *Bull. Soc. Fr. Min. Crist.* **92**, 160.
Zarka, A. (1974) *J. Appl. Cryst.* **7**, 453.

Appendix

Studies of natural diamonds
Cleavage plates
Ramachandran, G. N. (1944) *Proc. Indian Acad. Sci.* A19, 280.
Impurity platelets
Takagi, M. and Lang, A. R. (1964) *Proc. Roy. Soc.* A281, 310.
Origin of birefringence
Lang, A. R. (1967) *Nature,* **213**, 248.
Growth habit
Moore, M. and Lang, A. R. (1972) *Phil. Mag.* **26**, 1313.
Lang, A. R. (1974) *Proc. Roy. Soc.* **A340**, 233.
Abrasion of surfaces
Frank, F. C., Lawn, B. R., Lang, A. R., and Wilks, E. M. (1967) *Proc. Roy. Soc.* **A301**, 239.

Ion bombardment
Guseva, M. I., Ershova, L. M., Kiseleva, K. V., Krasnopevtsev,
          V. V., Milyutin, Yu. V. (1970) *Soviet Phys. Cryst.* 15, 441.
General review
Frank, F. C. and Lang, A. R. (1965) In *Physical Properties of Diamonds*
          (ed. Berman), p. 69, Oxford.

Study of mica
Willaime, C. and Authier, A. (1966) *Bull Soc. Fr. Min. Crist.* 89, 279.

CHAPTER 7

MELT, SOLID STATE AND VAPOUR GROWTH

7.1. MELT GROWTH

7.1.1. Semiconductors

It is unfortunate that X-ray topography was developed too late to particip-
ate in the first success at dislocation-free crystal growth. In 1959, with
X-ray topography still very much in its cradle, Dash (1959) succeeded in
pulling dislocation-free silicon crystals from the melt using Czochralski's
pulling technique (Fig. 7.1a). He characterized the crystals by his copper

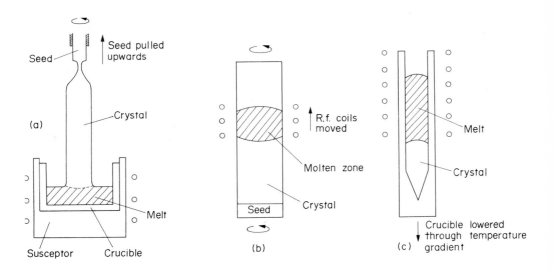

Fig. 7.1. (a) Schematic diagram of the Czochralski technique for
crystal growth from the melt. (b) Floating zone technique. (c) Bridgman technique.

decoration and infra red microscopy technique which can be used consistently
to reveal dislocation lines in silicon. Dash's method of growing a dislocation-free crystal relied on the production of a fine neck between the
seed and the growing crystal. Subsequent experiments using X-ray topography
have vividly shown how the dislocations grow out of the crystal in the region
of the neck enabling dislocation-free growth to take place when the crystal
diameter is enlarged. Of course, the crystal growth parameters must be
adjusted so that dislocations are not subsequently nucleated, but nowadays
sufficient expertise has been accumulated that crystal growers can repro-
ducibly grow dislocation-free silicon. Some manufacturers do, however, regularly employ topography as a perfection monitor by examining slices cut
from the top and bottom of the (large) crystals.

As indicated in section 4.1.4, current work on silicon is concerned with
the elimination of point defect clusters and growth striations. Patel and
Batterman (1963) observed a marked decrease in the intensity of anomalous
transmission of X-rays upon heat-treatment which was attributed to the precipitation of oxygen. Subsequently, Patel (1973, 1975) has used anomalous
transmission intensity measurements, section topography, and diffuse scattering measurements to investigate the effects of heat-treatment on silicon
crystals grown by a variety of methods. Czochralski-grown crystals regularly
contain a high concentration of oxygen, due mainly to contamination of the
melt by the crucible. Upon heat-treating, clustering of the point defects
occurs, and although no changes are observable in projection topographs there
is a very marked degradiation in the Pendellösung fringe pattern in the section topographs. Prior to heat-treatment, clear "perfect-crystal" interference fringes were seen (Fig. 7.2a) but as the heat-treatment progressed
(Fig. 7.2b) the pattern deteriorated. After 6 hours at 1000°C (Fig. 7.2c)
no Pendellösung fringes remained visible, and the defects were seen to have
condensed into bands. Under certain conditions the defects can be induced
to grow, and Authier and Patel (1975) have observed both faulted and prismatic loops. These findings have been confirmed by electron microscopy of
the smaller loops.

Use of the floating-zone technique for growth of silicon (Fig. 7.1b)
enables the oxygen and carbon concentrations to be drastically reduced.
Patel (1973) showed that when the oxygen content was reduced below present
detection levels, no reduction of intensity occurred on heat-treatment.
However, microdefects, which segregate into characteristic "swirl" patterns,
are often found in floating-zone crystals. Chikawa, Asaedo, and Fujimoto
(1970) have imaged the defects directly using a double crystal transmission
arrangement, de Kock (1974) and co-workers have revealed them by etching and
X-ray topography following lithium doping, and Patel (1973, 1975) has examined them by a combination of anomalous transmission and diffuse scattering
measurements. All experiments indicate the presence of defects less than
0.5 μm in diameter.

As mentioned in Chapter 4, growth under argon at high growth rates
enables "swirl" defects to be eliminated. However, upon increase in growth
rate there is an increase in the segregation of other impurities, notably,
carbon. This leads to flucutations in lattice parameter over the length of
the crystal (impurity striations). These striations have been studied in
some detail both in doped and undoped silicon crystals by de Kock, Roknoder,
and Boonen (1975) using a variant of section topography. They exploit the
high angular amplification in the centre of the Borrmann triangle to directly
image the carbon striations using high-order reflections for their section
topographs. It has been demonstrated that for crystals grown in the [111]
direction, the 444 reflection section topographs using MoKα and AgKα gave

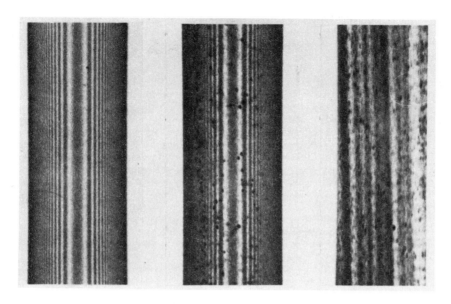

Fig. 7.2. X-ray section topographs of silicon containing appreciable
oxygen impurity showing the degradation of the Pendellösung
fringe pattern as point defect clusters grow during heat-
treatment. (a) As-grown, (b) after 1 hour at 1000°C,
(c) afer 6 hours at 1000°C (courtesy J.R. Patel).

comparable striation contrast to $\bar{8}80$ reflections in the (+-) parallel double
crystal setting. By use of a narrow diffracted beam slit, in a manner somewhat similar to the limited projection technique of Lang, de Kock *et al.* were
able to measure the amplitude of the striations in the centre of the Borrmann
triangle. Upon scanning of the crystal a quantitative measure of the fluctuations could be recorded via a photomultiplier and ratemeter. They found
that the carbon striations could be eliminated by subsequent annealing and
determined the conditions for striation-free growth.

An important study at present in progress in Japan is being performed by
Chikawa and collaborators. Using their X-ray video imaging system (Chikawa
et al., 1973) they have examined the growth of silicon *in situ* in the Lang
camera and observed the results of subsequent meltings and solidification.
Movement of a pair of heaters simulates growth by a floating zone. During
growth two types of dislocation are observed. One is stable and has a large
Burgers vector of <111>, and the other is unstable being the normal $\frac{1}{2}$<110>
slip dislocation. Upon melting dislocations moved away from the interface,
leaving that region almost dislocation free. Pendellösung fringes were
clearly visible. Conversely, on solidification, dislocations moved towards
the interface and definite evidence for dislocation generation behind the
interface was obtained.

Chikawa found that the common dislocations did not intersect the growth
interfaces while the stable <111> dislocations did so. When intersection
occurred, the facets disappeared in good agreement with the X-ray topographic

observations of Abe (1974) on the growth bands in bulk silicon. Abe found
that, during dislocation-free growth, the supercooling is greater than for
dislocated growth at the same rate. The supercooling was deduced from the
facet diameter in both these studies. The studies are far from complete and
only a little has yet been published. Some of the earlier work is reported
in Chikawa's (1974) review to ICCG-4 in Tokyo and the work described above
reported at the Limoges Summer school of 1975. The reader should watch the
literature as it is clear that with the power and flexibility of Chikawa's
system, which enables one to observe dislocation generation and movement
actually during growth, we may expect these experiments to contribute considerably to our understanding of melt growth.

   Although X-ray topography has been little applied to germanium either,
it has been important in characterization of the III-V compounds. Since the
work of Steinemann and Zimmerli (1967), Czochralski-grown gallium arsenide
has been produced with low, and sometimes zero, dislocation density. It is,
however, more difficult to completely eliminate dislocations as the high
vapour pressure of the arsenic makes liquid encapsulation necessary to avoid
unacceptable loss leading to non-stoichiometric composition. The use of
boric oxide as an encapsulent considerably effects the thermal conditions in
the crystal making dislocation-free growth difficult. Probably the most
important use of gallium arsenide is in light emitting diodes but not
insignificant is its potential as an injection laser. Much interest is
therefore centred around epitaxial layers of Ga-Aℓ-As-P on GaAs substrates.
Petroff and Hartmann (1973) have shown that in $Ga_{1-x}As_{1-y}Aℓ_xP_y$ layers the
presence of dislocations can terminate the operation of a laser. Therefore
it is vital to eliminate dislocations from the active region and this has
been achieved by Rozgonyi, Petroff, and Panish (1974). By careful control
of the stress introduced by the lattice mismatch it is possible to create a
unidirectional array of misfit dislocations at the interface which do not
propagate into the epilayer. Any dislocation from the substrate bows on
crossing the junction, and under suitable stress conditions it is possible to
make it run exactly parallel to the interface. As was observed on the topographs, no dislocations then penetrate the epilayer. The dislocation density
of the substrate was considerably less than that of the misfit dislocation
array - a feature attributed to multiplication of the misfit dislocations as
they glide in the plane of the interface out to the crystal surfaces.

7.1.2. Metals

   Until recently most metal crystals studied by topography were grown by the
Bridgman technique by slowly cooling the melt from one end in a pointed
mould (Fig. 7.1c). Provided the point is sufficiently sharp a single grain
nucleates at the tip and subsequently the crystal grows monocrystalline
unless growth conditions are so bad that heterogeneous nucleation takes place
during the solidification. Generally the dislocation density of Bridgman
metal crystals is high (10^6 cm^{-2}) due mainly to stresses exerted on the crystal by the mould on cooling. Young and Savage (1964) have shown that with
great care densities down to 10^4 cm^{-2} can be achieved. This is still rather
high for X-ray topographic study, and in order to reduce the density still
further a long anneal close to the melting point is required. It is found
that a cyclic anneal is helpful in preventing the formation of subgrain
boundaries (Kitajima, Ohta, and Tonda, 1974). With great care, particularly
in handling, slices may be prepared with sufficiently low densities for X-ray
studies. It is a tribute to the skill and patience of Young and Sherrill

(1967) that they have obtained such fine topographs of the early stages of
deformation of copper. They used neutron irradiation to pin the dislocations
during the cutting and polishing stages, and then annealed out the irradiation damage just prior to the topographic experiments.

Badrick and Puttick (1971) in their study of the growth of loops in cadmium took great care to control the cooling rate after annealing, but still
obtained quite high dislocation densities and large numbers of subgrains.

A basic problem to be overcome by the topographer is that it is extremely
difficult to cut slices from large single crystals of plastic material without introducing extraneous dislocations. Fehmer and Uelhoff (1969) have
developed an electrolytic-layer saw for cutting and polishing copper slices
for X-ray topography which does not introduce damage. Use of a spark erosion
machine leads to dislocation densities up to 10^4 cm^{-2} for distances up to a
few millimetres from the cut, and use of an acid saw gives a rough cut
although with little long range damage. Skilful use of the acid saw and subsequent chemical and electrochemical polishing is usually a satisfactory way
of preparing samples, but the sheer elegance and simplicity of Fehmer and
Uelhoff's device demands serious consideration for its adoption by any
topographer.

G'Sell, Champier, and Iwasaki (1974) overcame the problem by growing the
crystals as thin plates. Their method is simple and sketched in Fig. 7.3.

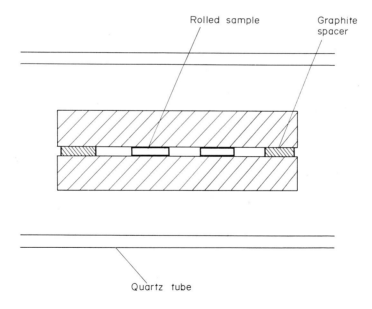

Fig. 7.3. Schematic diagram of the apparatus used by G'Sell *et al.*
to grow thin, highly perfect metal single crystals.

The crucible is horizontal, machined out of nuclear graphite, and consists of
two plates held a fraction of a millimetre apart by two pieces of graphite.
Rolled material is arranged between the plates, melted in an inert atmosphere,
and cooled very slowly. Somewhat surprisingly, the crystals are usually
monocrystalline and with low dislocation density. Fig. 7.4a shows a topograph
of cadmium crystal grown by this technique and oxidised for 76 hours. G'Sell
and Champier (1975) have studied the defects produced according to different
cooling rates and have observed some spectacular climb effects. Figure 7.4b
shows $^1/_3$<11$\bar{2}$3> type dislocations which, due to the presence of excess vacancies, have climbed into spirals. The movement is reversible as on heating
the spirals unwind. G'Sell and Champier have made some fine studies on hexagonal crystals grown by this method, and it has also been used to grow low
dislocation density silver and copper crystals (G'Sell, private communica-
tion). Here is another amazingly simple technique of which the X-ray topographer should take considerable note.

Experiments by Bonse, Kappler, and Uelhoff (1964) had shown that generally
metal crystals grown by the Czochralski technique were of superior quality to
those grown by the Bridgman technique. Although it has been appreciated only
recently, quite large crystals of metals can be grown dislocation-free by the
Czochralski technique. Howe and Elbaum (1961) many years ago reported that,
in conical "carrott shaped" crystals of aluminium, when the crystal diameter
became very small no dislocations were visible in the crystals. This recei-
ved little attention until Fehmer and Uelhoff (1972) and Sworn and Brown
(1972) independently reported growth of copper crystals free from dislocations
by pulling from the melt. With careful control of the pulling rate and the
melt cleanliness, and by growing a very thin neck between crystal and seed in
the manner pioneered by Dash, it is possible to obtain dislocation-free
growth. Fehmer and Uelhoff grew their crystals under vacuum and as the heat
losses at the specimen surface were only due to radiation, they were able to
grow much larger crystals than Sworn and Brown who grew under argon, and
therefore had additional heat losses due to convection. The former workers
sectioned their crystals prior at X-ray topographic analysis, but Sworn and
Brown used anomalous transmission topography to characterise their crystals
as-grown. Tanner (1973) used a similar technique to grow dislocation-free
silver crystals and an anomalous transmission topograph of such a crystal is
shown in Fig. 7.5. Screw dislocations parallel to the growth axis were not
observed, supporting the idea that growth occurs via a diffuse-interface
mechanism. Naramoto and Kamada (1974, 1975) have adapted the method to grow
dislocation-free niobium crystals by the floating-zone-pedestal technique.

While the relatively thin cylindrical Czochralski crystals are not very
suitable for X-ray topographic study because of the rapidly varying thickness,
they are ideal for preparing specimens for tensile testing. A parallel gauge
length and shoulders for mounting end-caps can be produced by simply varying
the temperature of the melt slightly. No cutting is required and it is
possible to mount the specimens in an Instrom tensile testing machine without
introducing dislocations. Kamada and Tanner (1974) performed a series of
tensile tests on copper crystals grown by this method and found marked differences in the mechanical properties of dislocated and dislocation-free
crystals. Anomalous transmission topography was used to check the specimen
perfection after mounting and we found that crystals between 1 and 1.5 mm in
diameter were amenable to both topography and tensile testing. Particularly
noteworthy among the results were the large and sharp yield drops observed
with dislocation-free crystals, similar to those found in whiskers. *In situ*
straining experiments in the Lang camera showed that slip started at one end
of the crystal and the Lüders band travelled down it as deformation

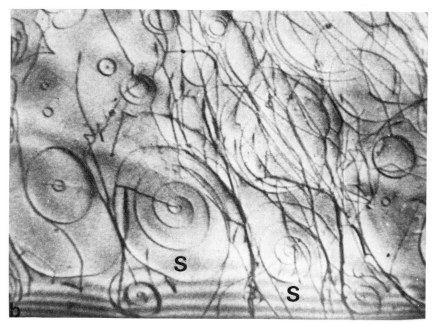

(Fig. 7.4)

Fig. 7.4. (a) Topograph of an oxidized cadmium crystal grown in
          the apparatus sketched in Fig. 7.3. (11$\bar{2}$0) reflection.
          Field width 2.3 mm. Note the loops L due to the oxida-
          tion. (b) Cadmium crystal cooled at 10°C per hour from
          187°C to room temperature. Spiral dislocations S provide
          clear evidence of extensive climb. (10$\bar{1}$1) reflection.
          Field width 3 mm (courtesy C. G'Sell).

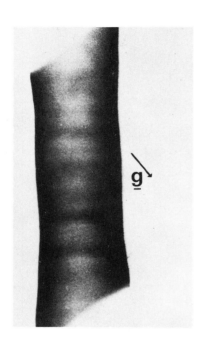

Fig. 7.5. Anomalous transmission
          topograph of a
          dislocation-free silver
          crystal grown by the
          Czochralski method.
          The crystal is circular
          in cross section and
          1.5 mm diameter.
          [110] growth direction
          is vertical. 111 ref-
          lection, MoKα radiation.

continued. However, on yielding the
rapid increase in dislocation density
destroyed the anomalous transmission
completely, and it was not possible to
follow the initial deformation stages.

7.1.3. Oxides

Many of the important oxides used
as laser materials or non-linear optical components are pulled from the
melt by the Czochralski technique.
Work by Sugii *et al.* (1973) on $LiNbO_3$,
which is important on account of its
non-linear optical coefficients, has
demonstrated that relatively low dislocation density crystals can be
pulled. They found that the perfection
was related to the growth direction –
crystals grown with [001] axis being
better than those with <210> axes.
Also they found that growth from a con-
gruent melt gave better crystals than
growth from a fresh stoichiometric
melt, and that sub-grain boundaries,
which cause considerable detrimental
light scattering, could thus be
avoided.
Earlier studies of Czochralski
grown Al_2O_3 showed that synthetic
sapphire grown by this method was of
superior perfection to that grown by
Verneuil's flame fusion technique.
May and Shah (1969) examined the
veiled region of a Czochralski sapphire
by topography, optical, and scanning
electron microscopy, and EPMA. They
found many cavities in the veiled
region which showed radial strain contrast in the topographs together with
dislocation networks. Generally in
oxide growth little care has been
taken to produce low dislocation
densities, as the optical perfection

has been the prime consideration. It is clear that there are many interesting features to be observed in currently available oxide crystals if

sufficient interest can be aroused.

Some time ago, Lang and Miuscov (1964) examined magnesium oxide crystals grown by slow cooling from the melt. They found well developed sub-grain structures, each sub-grain divided into several cells. Dislocations in the cell walls were individually resolved, the walls not being well aligned, and containing dislocations of several Burgers vectors. Large loops similar to those found in LiF were found and some crystals contained large helices indicating that a large amount of climb had taken place subsequent to solidification. In this respect, melt grown oxides show similar structures to metal crystals grown by solid state techniques. Finally, it was noted that many of the dislocations in MgO crystals were "dirty" and considerable precipitation had occurred along them. These dislocations correlated well with defects revealed by light scattering in the optical ultra-microscope (Lang and Miles, 1965).

7.1.4. Ice

Ice must seriously rival diamond as being the X-ray topographer's ideal material. It has low X-ray absorption, is cheap, and easy to grow. The early studies of Hayes and Webb (1965) were on dendritic ice crystals grown by supercooling distilled water to between -0.5 and -1.5° and seeding. They found that crystals were nearly dislocation-free as-grown but that subsequent handling led to dislocations being generated at the junctions of the dendritic branches. It was thought that inclusions between the branches were responsible for the dislocation generation. The topographs showed that the basal plane was the principal slip plane with dislocation Burgers vectors parallel to <$11\bar{2}0$> although some cross slip occurred.

Subsequent to their experiments on glacial ice crystals, Higashi, Oguro, and Fukuda (1968) grew ice crystals from the melt by a modified Czochralski technique and also in capillary tubes. They observed a marked asymmetry in the dislocation configuration as a function of growth axis orientation. The large Czochralski crystals had a dislocation density of about 10^4 cm^{-2}, the density being proportional to the growth rate but, significantly, the crystals contained no dislocations with axial screw components when growth was parallel to the c axis. In other directions axial screw components were found. Topographs of a thin crystal grown in a capillary tube parallel to the c axis showed it to be almost dislocation free. Beautiful Pendellösung fringes gave clear evidence of the perfection.

It appears that growth parallel to the c axis takes place by a two dimensional nucleation mechanism and this idea is substantiated by the mirror smooth surface of these crystals. The results agreed well with previous observations that the growth velocity perpendicular to the c axis varied as $v \propto (\Delta T)^2$, as predicted for spiral screw growth, whereas the growth velocity was considerably lower for growth parallel to the c axis.

Recently, Higashi and co-workers have developed a modified Bridgman method for growing ice crystals *in situ* in the Lang camera, thus enabling topographs to be taken during growth (Higashi, 1974). Although in crystals grown perpendicular to the c axis dislocations propagated through the neck grown between seed and crystal, when the interface was kept flat those dislocations ran parallel to the axis, and remained bunched together in the centre of the crystal. Crystals grown parallel to the c axis were nearly dislocation-free, no dislocations propagating from the seed. The thin neck between crystal and seed is again crucial to dislocation-free growth. The Japanese work on ice is reviewed by Higashi (1974).

7.2. SOLID STATE GROWTH

The first metal to be extensively studied by X-ray topography was aluminium grown by the strain-anneal method and during subsequent years considerable effort has gone into attempting to understand the mechanism of defect formation, notably by Nøst's group at Oslo and in Champier's laboratory at Nancy. In the work of Authier, Rodgers, and Lang (1965) topographs were taken of crystals cooled relatively rapidly. Well resolved dislocation helices were observed indicating that dislocation climb was very important in establishing the room temperature defect configuration.

Lohne and Nøst (1967) examined the formation of vacancy clusters and noted that during growth the clusters developed into a dislocation network. When clusters did not form, rows of dislocation loops were found. Nes and Nøst (1966) correlated the final dislocation density with the cooling rate and Nøst, Sørensen, and Nes (1967) showed that prismatic loops or rows of loops were nucleated at vacancy supersaturation or inverse subsaturation of about 11/10. Baudelet and Champier (1973) have extended the study of dislocation configurations during heat treatment, performing the heat treatments *in situ* whilst taking projection topographs. (The Norwegian work was confined to section topographs). They established under what conditions vacancy loops were nucleated and concluded that a Bardeen-Herring source mechanism for the nucleation of the rows of prismatic loops best fitted their observations. Following nucleation a loop expands by climb, and then loops glide away on their common glide cylinder to establish the equispaced row of loops.

Subsequent studies (e.g. G'Sell, Baudelet, and Champier, 1972) confirmed that most of the dislocation configurations observed in strain-anneal aluminium crystals could be described by dislocation climb due to condensation of vacancies. The work on aluminium is noteworthy in as much as it is one of the few systems to be studied systematically over an extended period of time by several independent workers. In view of the fine results which have emerged it must rank as one of the best applications of topography to date.

Several other materials have been successfully prepared for X-ray topographic examination by solid state techniques. Becker and Pegel (1969) observed long screw dislocations and helices parallel to <111> in strain-anneal grown molybdenum, again indicating that climb is a dominating effect in establishing the dislocation configuration. Brümmer and Alex (1970) grew crystals of tin by a press-temper-anneal method and found a low dislocation density in the products. Recently, Alex, Tikhonov, and Brümmer (1974) have observed magnetic domains in a nickel crystal grown by secondary recrystallization. An important advantage of strain-anneal or secondary recrystallization techniques is that the material emerges as a thin sheet which can either be used as it stands or else easily thinned for X-ray examination.

7.3. VAPOUR GROWTH

7.3.1. Whiskers

The mechanical properties of whiskers have excited much attention since the investigations of Brenner (1957). He showed that whiskers less than about 10 μm in diameter exhibited very high yield stresses, approaching those predicted for shear of a perfect crystal. As the thickness increases, the yield stress decreases, but a very sharp yield drop is still observed at the onset of plastic deformation (Yoshida, Gotoh, and Yamamoto, 1968). These

thick whiskers behave in a similar manner to dislocation-free copper crystals grown from the melt by the Czochralski method (Kamada and Tanner, 1974).
Consequently, the initial perfection of whiskers is of interest, both on
account of the mechanical properties and also because of the remarkable
morphology of these crystals. With regard to the latter, and it is quite
common for whiskers only a few microns in diameter to reach a centimetre in
length, it was long believed that whiskers contained an axial screw dislocation which promoted rapid unidirectional growth. However, the X-ray topographic studies of copper whiskers (Lyuttsau, Fishman, and Svetlov, 1966;
Nittono and Nagakura, 1969) and iron whiskers (Chikaura and Nagakura, 1972)
have all shown that an axial screw dislocation is not present in primary
whiskers. Very rarely, axial edge dislocations were observed by Lyuttsau
et al. and larger diameter whiskers were often imperfect and contained dislocations inclined to the growth axis. Nittono and Nagakura found that,
with careful growth by reduction of CuI with dry hydrogen, the whiskers up
to 80 μm which formed on the walls of the reaction boat were free from dislocations. They also observed that kinked and spiral whiskers were not only
monocrystalline but also dislocation-free. Although axial screw dislocations were observed in secondary copper whiskers (branched whiskers) less
than 0.1 μm thick by transmission electron microscopy (Hasiguti *et al*.,
(1970) there seems little doubt that growth of primary whiskers does not
take place via a spiral screw mechanism. Nittono has observed an increase
in the X-ray intensity at the whisker tip which is ascribed to the effect of
impurities, and this observation strongly supports the vapour-liquid-solid
model of whisker growth. On the VLS model a whisker grows with a blob of
liquid on the tip. The impurity concentration in the liquid will differ
from that in the solid and on final cooling, the enhanced impurity concentration will be frozen in at the tip.

   Chikaura and Nagakura also observed lattice distortions in iron whiskers
grown by reduction of ferrous chloride and subsequently annealed. The plate-
like whiskers showed distortions at the edges of the whisker and particularly at the roots of branching whiskers. Impurity, which can be removed by
annealing in vacuo or hydrogen, is known to be included in iron whiskers and
this strain is also attributed to the effect of impurity.

   An X-ray topographic study of the growth of whiskers *in situ* would be an
ideal method for clarifying the growth mechanism.

7.3.2. Metals

   Little work has been done on metals grown from the vapour, but it was
established some time ago that the perfection could be very high. Michell
and Smith (1968) grew cadmium and magnesium crystals, and the plate-like
crystals up to a millimetre across were of remarkably low dislocation
density. These crystals provided the first X-ray topographic observations
of dislocation movement during oxidation and paved the way for the later more
detailed studies (see section 4.1.3).

   Chikaura, Tomimatsu, and Nagakura (1974) have grown nickel plates by
reduction of nickel bromide in a stream of nitrogen and hydrogen. They
obtained crystals almost free from dislocations - the morphology of the crystal depending on the hydrogen content of the carrier gas. In the thicker
plates, images of ferro-magnetic domains have been observed.

7.3.3. Inorganic Crystals

Technologically, CdS is extremely important for its applications as a photo-device (the resistance is a function of illumination) and crystals are generally grown from the vapour. Chikawa (1967) used argon as a carrier gas with the starting material at one end of a silica tube at a temperature of $1150^{\circ}C$. The gas was streamed down the tube, and plate-like crystals grew at the cool end of the tube at about $950^{\circ}C$. The crystals were of very low dislocation density and Chikawa and Nakayama (1964) inferred that they grew by means of two types of dislocations which slipped out of the crystal after or during growth. In these crystals Chikawa observed Moiré fringes due to epitaxial overgrowths which had a slightly different lattice parameter due to inclusion of small quantities of impurity.

Thin vapour grown plates of layer compounds have been grown by iodine vapour transport techniques and have very low dislocation densities. Rimmington, Balchin and Tanner (1972) examined transition metal dichalcogenides grown by this method. Almost all the dislocations observed resulted from handling damage and perfect slip dislocations with Burgers vector parallel to <$11\overline{2}0$> were usually observed (Fig. 7.6.). In one instance, large area stacking faults bounded by partial dislocations were present. Although some of the crystals exhibited superb growth spirals on the surfaces, due to the fact that the crystals were very thin and as the spirals occurred on the large faces, the screw dislocation relating to the spiral could not be identified by X-ray topography. Buck (1973) has examined γ - In_2Sb_3 crystals grown by a very similar method and found dislocation arrangements much like those in the transition metal dichalcogenides. In a subsequent study of $ZnIn_2S_4$ and $Zn_3In_2S_6$ he observed dislocation reactions and noted that there was no correlation between defects and the point of attachment of the crystal to the crucible wall (Buck 1974).

Larger crystals grown from the vapour tend to be of lower crystalline perfection. For example, Al_2O_3 crystals of 20 mm length and 15 mm diameter examined by Farabaugh, Parker, and Armstrong (1973) using surface Berg-Barrett topography were found to contain subgrains, and individual dislocations were not

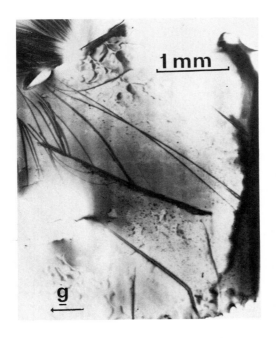

Fig. 7.6. Topograph of a crystal of the layer compound $SnS_{1.5}Se_{0.5}$ grown by iodine vapour transport. $01\overline{1}0$ reflection, MoKα radiation.

resolved.

Systematic investigations into the perfection of vapour grown crystals
as a function of growth conditions have not yet been performed and would be a
valuable contribution to our understanding of crystal growth mechanisms.

References

Abe, T. (1974) *J. Crystal Growth* 24/25, 463.
Alex, V., Tikhonov, L. V., and Brümmer, O. (1974) *Kristall und Technik* 9,
    643.
Authier, A. and Patel, J. R. (1975) *Phys. Stat. Sol. (a)* 27, 213.
Authier, A., Rogers, C. B., and Lang, A. R. (1965) *Phil. Mag.* 12, 547.
Badrick, A. S. T. and Puttick, K. E. (1971) *Phil. Mag.* 23, 585.
Baudelet, B. and Champier, G. (1973) *Crystal Lattice Defects* 4, 95.
Becker, C. and Pegel, B. (1969) *Phys. Stat. Sol.* 32, 443.
Bonse, U., Kappler, E., and Uelhoff, W. (1963) *Phys. Stat. Sol.* 3, K355.
Brenner, S. S. (1957) *J. Appl. Phys.* 28, 1023.
Brümmer, O. and Alex, V. (1970) *Phys. Stat. Sol. (a)* 3, 193.
Buck, P. (1973) *J. Appl. Cryst.* 6, 1.
Buck, P. (1974) *J. Crystal Growth* 22, 13.
Chikaura, Y. and Nagakura, S. (1972) *Jap. J. Appl. Phys.* 11, 158.
Chikaura, Y., Tomimatsu, M., and Nagakura, S. (1974) *J. Crystal Growth*
    24/25, 334.
Chikawa, J-I. (1967) *J. Phys. Chem. Solids* 28, Suppl. 1, 817.
Chikawa, J-I. (1974) *J. Crystal Growth* 24/25, 61.
Chikawa, J-I., Asaedo, Y., and Fujimoto, I. (1970) *J. Appl. Phys.* 41, 1922.
Chikawa, J-I., Fujimoto, I., Endo, S., and Mase, S. (1973) In *Semiconductor
    Silicon*, p. 448, Electrochem. Soc.
Chikawa, J-I. and Nakayama, T. (1964) *J. Appl. Phys.* 35, 2493.
Dash, W. C. (1959) *J. Appl. Phys.* 30, 459.
Farabaugh, E. N., Parker, H. S., and Armstrong, R. W. (1973) *J. Appl. Cryst.*
    6, 482.
Fehmer, H. and Uelhoff, W. (1969) *J. Phys. E* 2, 771.
Fehmer, H. and Uelhoff, W. (1972) *J. Crystal Growth* 13/14, 257.
G'Sell, C., Baudelet, B., and Champier, G. (1972) *J. Crystal Growth* 13/14,
    252.
G'Sell, C. and Champier, G. (1975) *Phil. Mag.* 32, 283.
G'Sell, C., Champier, G., and Iwasaki, H. (1974) *J. Crystal Growth* 24/25, 527.
Hasiguti, R. R., Yagi, E., Nishiike, U., and Sakai, T. (1970) *J. Crystal
    Growth* 7, 117.
Hayes, C. E. and Webb, W. W. (1965) *Science* 147, 44.
Higashi, A. (1974) *J. Crystal Growth* 24/25, 102.
Higashi, A., Oguro, M., and Fukuda, A. (1968) *J. Crystal Growth* 3/4, 728.
Howe, S. and Elbaum, C. (1961) *Phil. Mag.* 6, 1227.
Jenkinson, A. E. and Lang, A. R. (1962) In *Direct Observation of Imperfec-
    tions in Crystals* (ed. Newkirk and Wernick) p. 471, Wiley, New York.
Kamada, K. and Tanner, B. K. (1974) *Phil. Mag.* 29, 309.
Kitajima, S., Ohta, M., and Tonda, H. (1974) *J. Crystal Growth* 24/25, 521.
de Kock, A. J. R. (1974) *Philips Tech. Rev.* 34, 244.
de Kock, A. J. R., Roksnoder, P. J. and Boonen, P. G. T. (1975) *J. Crystal
    Growth* 28, 125.
Lang, A. R. and Miles, G. D. (1965) *J. Appl. Phys.* 36, 1803.
Lang, A. R. and Miuscov, V. F. (1964) *Phil. Mag.* 10, 263.
Lohne, O. and Nøst, B. (1967) *Phil. Mag.* 16, 341.

Lyuttsau, V. G., Fishman, Yu. M., and Svetlov, I. L. (1966) *Soviet Phys. Cryst.* **10**, 707.

May, C. A. and Shah, J. S. (1969) *J. Mater. Sci.* **4**, 179.

Michell, D. and Smith, A. P. (1968) *Phys. Stat. Sol.* **27**, 291.

Naramoto, H. and Kamada, K. (1974) *J. Crystal Growth* **24/25**, 531.

Naramoto, H. and Kamada, K. (1975) *J. Crystal Growth* **30**, 145.

Nes, E. and Nøst, B. (1966) *Phil. Mag.* **13**, 855.

Nittono, O. and Nagakura, S. (1969) *Jap. J. Appl. Phys.* **8**, 1180.

Nøst, B., Sørensen, G., and Nes, E. (1967) *J. Crystal Growth* **1**, 149.

Patel, J. R. (1973) *J. Appl. Phys.* **44**, 3903.

Patel, J. R. (1975) *J. Appl. Cryst.* **8**, 186.

Patel, J. R. and Batterman, B. W. (1963) *J. Appl. Phys.* **34**, 2716.

Petroff, P. M. and Hartmann, R. L. (1973) *Appl. Phys. Lett.* **23**, 469.

Rimmington, H. P., Balchin, A. A., and Tanner, B. K. (1972) *J. Crystal Growth* **15**, 51.

Rozgonyi, G. A., Petroff, P. M., and Panish, M. B. (1974) *Appl. Phys. Lett.* **24**, 251.

Steinemann, A. and Zimmerli, U. (1967) *J. Phys. Chem. Solids* **28**, Suppl. **1**. 81.

Sugii, K., Iwasaki, H., Miyazawa, S., and Niizeki, N. (1973) *J. Crystal Growth* **18**, 159.

Sworn, C. H. and Brown, T. E. (1972) *J. Crystal Growth* **15**, 195.

Tanner, B. K. (1973) *Z. Naturforschung* **28a**, 676.

Yoshida, K., Gotoh, Y., Yamamoto, M. (1968) *J. Phys. Soc. Japan* **24**, 1099.

Young, F. W. Jr. and Savage, J. R. (1964) *J. Appl. Phys.* **35**, 1917.

Young, F. W. Jr. and Sherrill, F. A. (1967) *Canadian J. Phys.* **45**, 757.

Appendix

Melt Growth

Silicon

de Kock, A. J. R. (1970) *Appl. Phys. Lett.* **16**, 100.

de Kock, A. J. R., Beeftinck, F. M., and Schell, K. J. (1972) *Appl. Phys. Lett.* **20**, 81.

Yoshimatsu, M. (1963) *J. Phys. Soc. Japan* **18**, Suppl. 2, 335.

Yukimoto, Y. (1968) *Jap. J. Appl. Phys.* **7**, 348.

Copper

Wittels, M. C., Sherrill, F. A., and Young, F. W. Jr. (1962) *Appl. Phys. Lett.* **1**, 22.

Young, F. W. Jr., Baldwin, T. O., Merlini, A. C., and Sherrill, E. A. (1966) *Adv. X-ray Analysis (Plenum)* **9**, 1.

Young, F. W. Jr. (1967) *J. Phys. Chem. Solids* **28**, Suppl. 1, 789.

Clareborough, L. M., Michell, D., and Smith, A. P. (1967) *Phys. Stat. Sol.* **21**, 369.

Tanner, B. K. (1972) *J. Crystal Growth* **16**, 86.

Basariya, A. G., Kapanadaze, B. K., and Sanadze, V. V. (1973) *Soviet Phys. Cryst.* **18**, 411.

Kuriyama, M., Early, G., and Burdette, H. E. (1974) *J. Appl. Cryst.* **7**, 535.

Gallium

McFarlane, S. H. III and Elbaum, C. (1967) *J. Appl. Phys.* **38**, 2024.

Garnets

Glass, H. (1972) *Mater. Res. Bull.* **7**, 385, 1087.

Blank, S. L. and Nielsen, J. W. (1972) *J. Crystal Growth* **17**, 302.

Ice

Oguro, M. and Higashi, A. (1971) *Phil. Mag.* **24**, 713.

Parafin
Green, R. E. Jr., Farabaugh, E. M., and Crissman, J. M. (1975) *J. Appl.
          Phys.* <u>46</u>, 4173

Solid State Growth
Aluminium
Nøst, B. (1965) *Phil. Mag.* <u>11</u>, 183.
Nøst, B. and Sørensen, G. (1966) *Phil. Mag.* <u>13</u>, 1075.
G'Sell, C., Baudelet, B., and Champier, G. (1972) *J. Crystal Growth* <u>13/14</u>,
          252.
Fremiot, M., Baudelet, B., and Champier, G. (1968) *J. Crystal Growth* <u>3/4</u>,
          711.
Titanium
Jourdan, C., Rome-Talbot, D., and Gastaldi, J. (1972) *Phil. Mag.* <u>26</u>, 1053.

Vapour Growth
Epitaxial silicon
Schwuttke, G. H. and Sils, V. (1963) *J. Appl. Phys.* <u>34</u>, 3127.
Schwuttke, G. H. (1966) *Trans. 3rd Int. Vacuum Congress* <u>2</u>, 301.
Epitaxial gallium phosphide and gallium arsenide
McFarlane, S. H. III and Wang, C. C. (1972) *J. Appl. Phys.* <u>43</u>, 1724.
Tellurium
Naukkarinen, K. and Tuomi, T. O. (1969) *J. Appl. Phys.* <u>40</u>, 3054.
Epitaxial silicon carbide
Isherwood, B. J. and Wallace, C. A. (1965) *J. Appl. Cryst.* <u>1</u>, 145.
Copper-gold alloy whisker
Nittono, O., Onodera, N., and Nagakura, S. (1970) *Jap. J. Appl. Phys.* <u>9</u>, 328.
Cadmium sulphide
Chikawa, J-I. (1967) *Adv. X-ray Analysis (Plenum)* <u>10</u>, 153.
Cadmium selenide
Skorokhod, M. Ya. and Datsenko, L. I. (1968) *Soviet Phys. Cryst.* <u>13</u>, 459.

INDEX

Absorption coefficient,7,8,75,130,139.
Absorption topography,148.
Acid saw,159.
Alpha particle channelling,122-3.
Aluminium,8,77,103,126,160,164,169.
Aluminium oxide,166.
　　(see ruby and sapphire)
Amethyst,148.
Ammonium dihydrogen phosphate,144.
Ammonium hydrogen oxalate,134-5.
Angular amplification,70,76,156.
Anomalous dispersion,116-7.
Anomalous transmission,6ff,33,40,75,
　　102,140-3,156,160,162.
Antiferromagnets,119,121,128.
Apatite,152.
Arsenic,126.
Asymmetric reflection,21ff,27,51,122.
Avogadro's number,56.

Bardeen-Herring source,164.
Barite,152.
Barium titanate,116-8,128,139.
Barth-Hoseman technique,26,27ff.
Bend contours,97.
Benzil,134.
Berg-Barrett method,24,25ff,42,61,98,
　　102,119,166.
Beryllium oxide,116,139,143.
Bimodal profile of dislocations,83.
Birefringence,14,117,119,148,150,153.
Bitter technique,119.
Bloch waves,2ff.
Boron,84,107.
Borrmann effect (see anomalous
　　transmission)
Borrmann fan,11,14,23,28,64,67,70,78,
　　81-2,88,156-7.
Borrmann triangle (see Borrmann fan)
Bragg angle controller,35.
Brass,126.
Brazil twin,116,148-9.
Bridgman technique,155,158,160,163.
Brillouin zone boundary,14.
Burgers vector analysis,57,77,79,83,
　　100ff,135-6.
Buried layer,122.
Butterfly contrast,119,121.

Cadmium,159,162,165.
Cadmium selenide,169.
Cadmium sulphide,53,166,169.
Calcite,103,116,138,146.
Channel plate,46.
Chromium,119.
Climb,104,107,162,164.
Cobalt oxide,119,128.
Cobble texture,136.
Coercive field,117.
Coercivity,119.
Collimators,30,51.
Column approximation,78.
Columnar growth,146,148.
Copper,102-3,127,129,159,160,165,168.
Copper-gold whisker,169.
Copper whisker,126,165.
Crystal mounting,32.
Curved crystal technique,42.
Cyclotrimethylene trimitramine,144.
Czochralski technique,155-6,158-63,
　　165.

Dauphiné twin,116,148.
Debye-Waller factor,8.
Dendrites,163.
Dental film,38,46.
Detection efficiency,45.
Deviation parameter,11,18,93,115.
Device fabrication,84,86,104ff,122.
Diamond,69,114,116,145-8,151,153.
Diffracting zone model,82.
Diffuse scattering,156.
Diffusion induced defects,104ff,127.
Diodes,108.
Direct image,28,31,69,77,81-3,96.
Disclination,119.
Dislocation dipoles,104,107.
Dislocation-free crystals,50,109,130,
　　138,139,155-60,163,165.
Dislocation velocity,103-4,127.
Dispersion,50.
Dispersion surface,4ff,64,70,72-4,77.
Domains
　　-Ferroelectric,116-8,128,140.
　　-Magnetic,27,59,93-5,99,119-22,
　　　　　　　　128,140,164-5.
Dolomite,150.
Double crystal method,24,47ff,56-7,
　　61,101,111-3,156-7.
Double crystal spectrometer,15,17.

DuMond diagram,48.
Dynamical image,28,39,69,81,121.
Dynamical theory,2ff.

Effective lens,51.
Eikonal,71,74-6,78,80,98.
Elastic constants,134-5.
Electrolytic layer saw,159.
Electron probe analysis,136,162.
Emitter-edge dislocations,108.
Etch pits,102,150.
Evanescent waves,9.
Ewald sphere,3,34,63.
Exposure times,40ff,43.
Extinction contrast,24,59,116.
Extinction correction,2,130.
Extinction distance,17,37,65,74,82,93,
    116,131-2.

Facets,112-3,157-8.
Faraday effect,111,119.
Ferrite,128.
Flame fusion technique,162.
Floating zone technique,155-7,160.
Fluorite,150-1.
Frank dislocations,114.
Frank-Read source,102,103.
Fresnel zone,64.
Friedel's Law,29,131.

Gadolinium gallium garnet,86,111-3.
Gadolinium molybdate,126,128.
Gallium,168.
Gallium arsenide,53,158,169.
Gallium phosphide,126,169.
Garnet,93,111,128,139,168.
Gemstones,145.
Germanium,52,67,126,127,135,158.
Grown-in dislocations,103.
Growth bands,50,95,112-3,133,138,140-1,
    145-6,148-52,156.
Growth sector,130,134,136,147-50.
Growth sector boundaries,137,141,148,
    150-1.
Growth striations (see growth bands).
Hafnium diboride,144.
Heidenreich-Shockley faults,114.
Helical dislocations,103,112,139,163-4.
Hexamethylene tetramine,130,133-4,144.
Hook-shaped fringes,67,69.
Horizontal stripes,34.
Hot margin,69-70.

Hot pouring of flux,140-1.
Hour-glass pattern,88-9.

Ice,104,114,126,127,152,163,168.
Image doubling,26,28,37.
Image intensifier,47.
Image orthicon,47.
Incidence plane,50,87.
Indium antimonide,126.
Indium phosphide,126.
Indium sulphide,166.
Injection laser,158.
Integrated circuits,104ff.
Inter-branch scattering,78,95,101,136.
Interferometer,55-7,61,99,122.
Intermediary image,81-4,101.
Ion implantation,110,122-3,129.
Iron,94,104,126,127-8.
Iron borate,140.
Iron-silicon,27,93-4,102,119-22,128.
Iron whisker,119,128,165.
Isoberyl,144.

KDP (see potassium dihydrogen
    phosphate).
Kerr effect,119.
Kinematical theory,1.

Lang camera,29,62,103,157,163.
Lang technique,24,28ff,42,52,57,59,61,
    93,99.
Lattice parameter measurement,50,53,
    56,111-3.
Laue equation,2,4.
Laue pattern,58.
Laue point,4.
Layer compounds,166.
Lead germanate,116.
Limited projection technique,31,157.
Lithium decoration,111.
Lithium fluoride,163.
Lithium formate,134.
Lithium niobate,128,162.
Lomer dislocation,102.
LOPEX silicon,50.
Lüders band,160.

Magnesite,150.
Magnesium,165.
Magnesium oxide,150,163.
Magnetic bubbles,104,111.

Magnetostriction193,119.
Margin effect (see hot margin).
Maxwell's equations,2,20.
Mica,152,154.
Microscopes,40.
Moiré fringes,53ff,131-2,166.
Moiré sandwich,53-4.
Molybdenum,103,164.
Monochromators,18,21.
Mosaic model,83,102.
Multiple crystal arrangements,61-2.

Neutron interferometer,56.
Neutron irradiation,159.
Neutron topography,62,67,119.
Nickel,164,165.
Nickel oxide,93,119,128.
Niobium,160.
Nuclear emulsions,38,44,46,59.

OMD method,52.
Orientation contrast,24,59,93,116.
Organic crystals,130,134-5,144.
Orthoferrite,119,128,144.
Oxalic acid,130.
Oxidation,104-6,127,162,165.

Paraffin,169.
Peierels energy,135.
Pendellösung,14ff,82,98.
Pendellösung fringes,63-70,88,93,130,
    131-3,140,156-7,163.
    -in distorted crystals,98.
    -visibility,65.
Penning-Polder theory,71ff,84,93,96,
    101.
Permitivity,117.
Photography,38.
Plastic deformation,108,127,149,150,
    160,164.
Poisson's ratio,100.
Polarization,5,66-7,69.
Potash alum,130,145.
Potassium dihydrogen phosphate,126,
    134,144.
Potassium cobalt fluoride,119,122.
Potassium nickel fluoride,50,119,121-2,
    140,144.
Poynting vector,10,73.
Precipitates,75-7,84,140-3,150,163.
Projection topograph,28ff,65,82-3,
    89-92,100,156,165.

Pseudo-plane wave,14,17.

Quartz,53-4,91,104,116,128,136-8,
    142,145,148-50.
Quasi-dislocation theory,119.

Radiation damage,70,129,159.
Radiation hazard,27-8,30.
Rare earth salts,58,76,86,116,140-3.
Ray path parameter,74.
Refraction of dislocations,134,148,
    151-2.
Refractive index,56.
Resolution,27,34,35ff,44-5,56,110.
Reviews
    -applications,100.
    -contrast of dislocations,81.
    -diamond,145,154.
    -dynamical theory,2.
    -ice,163.
    -interferometry,56.
    -semiconductors,110,111.
    -spherical-wave theory,67.
    -techniques,24,27.
    -transducers (modes of oscilla-
    tion),104.
    -video topography,44.
Rocking curve,18,26,34,37,48,63,93,
    101.
Rowland circle,52.
Ruby,139,140,146.

Salol,133.
Sapphire,140,144,162.
Scanning electron microscope,162.
Schmidt factor,103.
Schrodinger equation,3.
Schulz method,25,56.
Screw dislocation growth,138,139,
    145,152,163,165-6.
Secondary recrystallization,164.
Selenium,126.
Section topograph,28ff,65-71,77-82,
    84-9,91-3,101,114,132,137,156-7,
    164.
Shockley faults,114.
Shot noise,39.
Silicon,8,50,53,56-7,63-4,67,70,77,
    84,86,91-2,102-3,107-111,114,122,
    126,127,129,135,155-8,168-9.
Silicon carbide,169.
Silicon diode array,44,111.

Silicon web dendrite,114-5.
Silver,160.
Silver iodide,144.
Simulated images,79,92,101.
Sirtl etch,111.
Sodium chlorate,131,134.
Sodium chloride,130-2,134.
Sodium nitrite,117.
Soller slit,28.
Spark erosion,159.
Spike reflection,146.
Spin density wave domains,119.
Spinel,144.
Spiral dislocations,162.
Spontaneous polarization,116.
Stacking fault,21,69,86ff,99, 114-6,
   136,146.
Stacking fault tetrahedra,114,146.
Stage design,61.
Stair-rod dislocations,116.
Stepper motor,32,52.
Stereo pairs,29,101-2.
Strain-anneal growth,164.
Structure factor contrast,117.
Structure factor measurement,51,65.
Sulphur,144.
Surface damage,70,86,136.
Surface skin,137,140.
Susceptibility,2,7.
Swirl,110,156.
Synchrotron radiation,25,44-5,56-9,
   103.
Syton,86.

Takagi's equations,20-1,79-81.
Tellurium,169.
Thickness fringes,122.(see also
   Pendellösung fringes).
Thiourea,134.
Tie point,4,9,12,18,22,64,67-8,73,78,
   88,93-5.
Tie point jumping (see inter-branch
   scattering).
Tie point migration173ff.
Tin,105-6,164.
Tin disulphide,166.
Titanium,169.
Titanium diboride,144.
Topaz,151-2.
Transistors,108.
Translation fault,122.
Transmission electron microscopy,64-5,
   78,165.
Traverse striations,29.

Traverse topograph (see projection
   topograph).
Triglycine sulphate (TGS),117,129,
   136,144.
Trigons,148.
Twin,89-90,93-5,116,139,148-50.

Vacancy clusters,111,164.
Vapour-liquid-solid growth,165.
Vibration,35.
Video topography,44,103,110,157.
Visibility,45.

Whiskers,119,126,128,160,164-5,169.

X-ray line width,36.

Yield drop,160,164.

Zebra stripes,50.
Zener diode,109.
Zinc,104,127.
Zinc oxide,138.
Zinc sulphide,126.
Zirconium diboride,144.